THE MORPHOSTRUCTURE OF THE ATLANTIC OCEAN FLOOR

OCEANOGRAPHIC SCIENCES LIBRARY

V. M. LITVIN

P. P. Shirshov Institute of Oceanography, Moscow, U.S.S.R.

The Morphostructure of the Atlantic Ocean Floor

Its Development in the Meso-Cenozoic

Translated from the Russian by

V. M. DIVID, N. N. PROTSENKO and YU. U. RODZHABOV

D. REIDEL PUBLISHING COMPANY

A MEMBER OF THE KLUWER ACADEMIC PUBLISHERS GROUP

Dordrecht / Boston / Lancaster

Library of Congress Cataloging in Publication Data

CIP

Litvin, V. M. (Vladimir Mikhaĭlovich)
The morphostructure of the Atlantic Ocean floor, its development in the Meso-Cenozoic.

(Oceanographic sciences library)
Translation of: Morfostruktura dna Atlanticheskogo okeana i ee razvitie v Mezozoe i Kaĭnozoe.
Bibliography: p.
Includes index.
1. Ocean bottom–Atlantic Ocean. 2. Geology, Stratigraphic–Mesozoic.
3. Geology, Stratigraphic–Cenozoic. I. Title. II. Series.
GC87.2.A86L5713 1984 551.46′08′093 83-26908
ISBN 90-277-1509-2

Translated from the 1980 edition of
Морфоструктура дна Атлантического океана и ее развитие в мезозое и кайнозое
published by Nauka, Moscow, U.S.S.R.

Published by D. Reidel Publishing Company,
P.O. Box 17, 3300 AA Dordrecht, Holland

Sold and distributed in the U.S.A. and Canada
by Kluwer Academic Publishers
190 Old Derby Street, Hingham, MA 02043, U.S.A.

In all other countries, sold and distributed
by Kluwer Academic Publishers Group,
P.O. Box 322, 3300 AH Dordrecht, Holland

Printed in The Netherlands

Contents

Introduction

The study of the topography and structure of the ocean floor is one of the most important stages in ascertaining the geological structure and history of development of the Earth's oceanic crust. This, in its turn, provides a means for purposeful, scientifically-substantiated prospecting, exploration and development of the mineral resources of the ocean.

The Atlantic Ocean has been geologically and geophysically studied to a great extent and many years of investigating its floor have revealed the laws governing the structure of the major forms of its submarine relief (e.g., the continental shelf, the continental slope, the transition zones, the ocean bed, and the Mid-Oceanic Ridge). The basic features of the Earth's oceanic crust structure, anomalous geophysical fields, and the thickness and structure of its sedimentary cover have also been studied. Based on the investigations of the Atlantic Ocean floor and its surrounding continents, the presently prevalent concept of new global tectonics has appeared. A great number of works devoted to the results of geomorphological, geological, and geophysical studies of the Atlantic Ocean floor have appeared. In the U.S.S.R., such summarizing works as *The Geomorphology of the Atlantic Ocean Floor* [34], *Types of Bottom Sediments of the Atlantic Ocean* [24], *The Geology of the Atlantic Ocean* [38], and, somewhat earlier, *Geophysical Studies of the Earth's Crust Structure in the Atlantic Ocean* [13], have been published.

It should be stressed, however, that each of the above-mentioned publications follows a specific trend and deals mainly with one particular aspect of ocean-floor studies. In these works, morphostructural anslysis aimed at ascertaining the interrelations between the submarine relief and the geological structure of the oceanic crust, determining the role of tectonic movements, volcanism, and sedimentation in the formation of structural elements of the ocean floor relief and verifying the history of its development, is either not considered at all or used to a very small extent. It is well known that the methods of morphostructural analysis and the morphostructural approach in land studies were developed long ago and have now found extensive application [15]. In ocean-floor studies, however, this method, up to now, has been used only occasionally. The main objective of our work in studying the Atlantic Ocean floor was, therefore, to develop the morphostructural approach in submarine geomorphological, geological and geophysical investigations. This work was carried out over a period of several years in the Laboratory of Atlantic Geology at the Atlantic division of the P. P. Shirshov Institute of Oceanology, U.S.S.R. Academy of Sciences, and was done within the framework of the International Geodynamics Project.

The specific objectives of the work were: (1) to reveal the interrelations between the submarine relief and the geological structure of the Atlantic Ocean floor, between the structural forms of the relief and anomalous geophysical fields, between the structure of the sedimentary sequence and the relief of the oceanic basement; (2) to determine the role of horizontal and vertical tectonic movements, faults, volcanism and sedimentation in the formation of ocean-floor morphostructure; (3) to ascertain the laws governing the spatial and temporal development of the morphostructural and morphotectonic zones of the ocean in the Meso-Cenozoic; to clarify the history of ocean-floor development in terms of the concept of global plate tectonics.

The methodological basis of research was laid down by morphostructural analysis, known to include the use, comparison and integrated interpretation of geomorphological, geological and geophysical data. The concept of global-plate tectonics and its application to the development of the ocean-floor morphostructure in the Meso-Cenozoic served as the theoretical basis of the work. At the same time, the possibility was considered of applying the concept of large-scale subsidences of the continental crust and its reworking into the crust of an intermediate or sub-oceanic type ('oceanization') to explain the origin of the continental margins and the adjoining parts of the Atlantic Ocean floor.

The morphostructure (and in a more general concept, the morphotectonics) of the ocean floor was chosen by the author as the main object of study. Morphostructure is understood as the manifestation of the ocean floor's geological structure in a submarine relief. The same meaning, but in a more general sense, is given to the concept of morphotectonics. Large and medium-size forms of submarine relief, whose formation is caused by tectonic movements and geological structure, are regarded as specific morphostructures. In the present work, while analysing the data on the structure and genesis of the relief and structural characteristics of the ocean floor, the primary emphasis is placed on the influence of endogenous factors, with sedimentation being the only exogenous factor considered.

The work done was based on the data of geomorphological, geological and geophysical studies conducted on board the research vessels of the Institute of Oceanology, *Akademik Kurchatov* and *Dmitri Mendeleev*, as well as on other Soviet research vessels, such as *Mikhail Lomonosov, Akademik Vernadsky, The Pole, Sevastopol, Akademik Knipovich, Petr Lebedev* and others. Over the period of 1967–1977 *Akademik Kurchatov* made 17 voyages in the Atlantic (including five specialized geological and geophysical voyages) during which geomorphological, geological and geophysical studies were made. Altogether during these voyages, 250 000 miles of continuous echo-soundings, more than 100 000 miles of magnetic survey and 30 000 miles of continuous seismic profiling (CSP) were performed. Twenty-four polygons, 16 of them geophysical, were run with comprehensive geological and geophysical studies and echo-sounding and some geological or geophysical work was performed on the others (Figure 1).

We have also used the echo-sounding data of other Soviet expeditions and the data of other expeditions have been resorted to. By the beginning of 1976, numerous expeditions, mainly American, had performed more than 400 000 miles of continuous seismic profiling. Regional magnetometric, gravimetric, and seismic investigations have been carried out and a large number of bedrock and bottom sediment samples have been studied. Of great importance for the understanding of ocean-floor structure are the deep-sea drilling data obtained on board R/V *Glomar Challenger*. Over the period of 1968–1976 more than 160 boreholes were drilled in the Atlantic Ocean and the Norwegian-Greenland Basin, and the data thus obtained have been extensively used in this work.

Based on the analysis of the above data, the author has described the general scheme of the ocean-floor relief and the geomorphological features of individual morphostrcutures. The data of seismic profiling and deep-sea drilling pertaining to the structure of sedimentary cover have been summarized. The geophysical data on the Earth's crust structure and anomalous geophysical fields have been correlated with the morphostructure of the ocean floor. Seismotectonics, volcanism and ocean-floor faults have been studied, and their role in submarine relief formation ascertained. The role of horizontal and vertical tectonic movements in the formation of ocean-floor morphostructure has also been assessed, and this served as the basis for the elaboration of a new scheme of its development in the Meso-Cenozoic.

The author has compiled a number of new maps of the Atlantic Ocean – physiographic, geomorphological, sedimentary cover thickness, seismotectonic, morphotectonic, horizontal and vertical tectonic movements, as well as the paleomorphostructural schemes for different stages of ocean development in the Meso-Cenozoic.

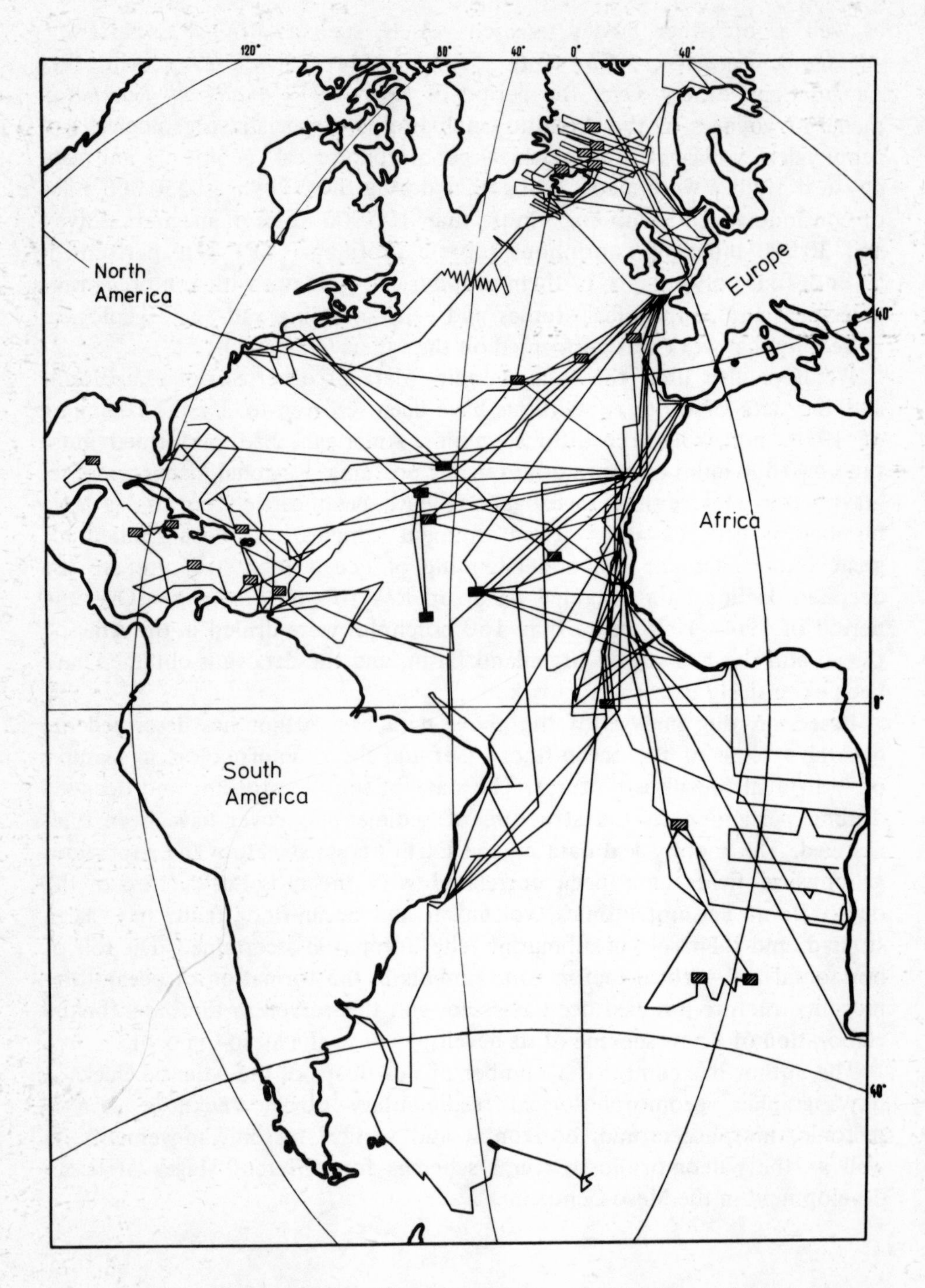

Fig. 1. Map of the expeditions undertaken by the Institute of Oceanology, U.S.S.R. Academy of Sciences, in 1967–1976.

Abstract

This monograph summarizes the results of geomorphological, geological and geophysical studies that have been conducted in the Atlantic Ocean over a period of many years on board the R/V '*Akademik Kurchatov*' and by other expeditions, the results of deep-sea drilling, and numerous published data on the ocean floor structure. The present work has been carried out in accordance with the programme of the International Geodynamic Project. Morphostructural analysis served as its methodological basis, and the concept of global plate tectonics as its theoretical basis. At the same time, the development of continental margins and the adjacent parts of the ocean was examined in its association with large-scale submersions of the Earth's crust and possible 'oceanization'. A number of new maps of the ocean floor have been compiled, reflecting its structure, the role of various factors in its formation, and the ocean floor morphostructure development in the Meso-Cenozoic.

The general scheme of the structure of ocean floor relief and the geomorphological peculiarities of individual morphostructures are described. The features of symmetry and circumcontinental zonality in the submarine relief, explained by ocean floor spreading, are revealed. Sedimentary cover thickness and age are observed to increase on both sides of the Mid-Atlantic Ridge axis in the direction towards the continental margins. The interrelations of the oceanic basement relief and that of the ocean floor are shown. The role of sedimentation in relief formation, consisting in the shaping of morphostructures created by endogenous processes, is estimated. The influence of sedimentation, up to complete smoothing-out of the primary relief of the basement, is seen to manifest itself to the greatest extent on continental margins, in foredeeps and the peripheral areas of oceanic basins.

The Earth's crust structure, anomalous geophysical fields (magnetic, gravitational, thermal) and their relationship with the ocean floor morphostructure are described. A marked difference is noted between the structure of the Earth's crust on continental margins and on the ocean floor, indicating that they were formed by different processes. Special attention is given to banded magnetic anomalies characteristic of the oceanic crust and indicative of ocean floor spreading.

The data on ocean floor seismicity, volcanism and fractures are presented. Seismological data indicate tension in the rift zone and compression in the zone of deep-sea trenches. Volcanic processes play the decisive role in forming the oceanic basement and volcano-tectonic morphostructures of the ocean floor, and also participate directly in creating seamounts and volcanic islands. Faults on the ocean floor, the latitudinal ones being decisive, are indicative of a layered-block structure of the Earth's crust in the ocean.

The role of tectonic movements in the formation of ocean floor morphostructure is assessed. Horizontal movements, of which the most significant are the processes of lithospheric plates spreading away from the axial rift fault on both sides of it, determine the relative positions of individual ocean floor morphostructures and the overall morphostructural plan. An important role is also played by the regional movements of individual blocks, e.g., in the Caribbean and the South Antillean Transition Zones. Vertical movements, taking place simultaneously with the horizontal ones and interconnected with them, determine the formation of the morphostructures themselves, their height and disjunction. In the course of spreading the lithospheric plates gradually subsided, and as a result the Mid-Atlantic Ridge and the oceanic basins on both sides of it appeared. During the Meso-Cenozoic the subsidence was practically uninterrupted on continental margins, leading to sedimentary cover accumulation and the creation of epicontinental platforms. Transition zones are characterized by a complicated combination of the rising ridges of island arcs and the submerging basins of marginal seas.

Three main stages are distinguished in the development of the Atlantic Ocean floor morphostructure: (a) the stage of the opening of the ocean (Late Jurassic–Early Cretaceous), (b) the stage of the formation of basic morphostructures (Late Cretaceous–Early Paleogene), (c) the neotectonic stage (Late Paleogene–present-day period). At the first stage, as a result of the break-up of the primeval continent, a narrow basin (or a series of basins) appeared, from which the Atlantic Ocean was later created. At the second stage all the basic morphostructures were formed, such as the Mid-Oceanic Ridge, oceanic basins, dome-and-block uplifts, transition zones, continental foredeeps and epicontinental platforms. At the neotectonic stage the formation of these morphostructures and their shaping by exogenous processes, the most powerful of which was sedimentation, were completed. Formation of new morphostructures in the rift zone of the Mid-Oceanic Ridge and in the areas of the island arcs of transition zones continued.

CHAPTER 1

Bathyography of the Atlantic Ocean Floor

1. PRINCIPAL FORMS OF RELIEF AND GEOMORPHOLOGICAL ZONING OF THE OCEAN FLOOR

In the present work, the Norwegian-Greenland Basin is considered together with the Atlantic Ocean. Inland seas, such as the Baltic, Mediterranean and Black seas are not dealt with as their structure is not directly associated with the ocean.

A characteristic feature of the Atlantic Ocean floor relief is the Mid-Oceanic Ridge stretching from north to south approximately midway between the coasts of Europe and North America and Africa and South America. The ridge axis bends following the general conformation of the ocean and the coastlines. In the south the ridge bends round Africa and continues in the Indian Ocean. In the north it passes through Iceland and the Norwegian-Greenland Basin continuing in the Arctic Basin. The Mid-Atlantic Ridge is part of the global system of mid-oceanic ridges encircling the Earth's ocean floor.

On both sides of this grandiose structure are located the abyssal basins of the Atlantic Ocean separated either by transverse (relative to the Mid-Oceanic Ridge strike) rises and ridges or by elevated zones of hilly relief. The depth of these depressions reaches 3–6 km. The floor is comprised of flat and hilly plains with elevations, plateaus, hills and seamounts. Along the coasts of its surrounding continents stretch the shelf zones ending on their seaward side in escarpments of the continental slopes. The structure and geomorphology of the underwater marginal parts of the continents are retained in the shelves. The bordering continental slopes are, in essence, the flanks of continental blocks and represent the zones of transition between the shelf and the ocean bed. In places, the continental slopes are complicated by banks located at different depths and usually called marginal plateaus.

A special place in the ocean-floor relief is occupied by complex transition zones comprising island arcs, deep-sea trenches conjugated with them, and the depressions of marginal seas separated by them from the ocean.

Such a type is known to be widespread in the western part of the Pacific Ocean, but in the Atlantic Ocean it is only found in two regions: the Caribbean and the Scotia Seas.

Almost everywhere at the foot of the continental slopes are located inclined plains, formed by accumulated sediments, which are called continental rises. Although morphologically they form the lower parts of the continental slopes, structurally they should be regarded as forms superimposed on the marginal parts of the ocean floor.

The following major forms (provinces) of the submarine relief on the Atlantic Ocean floor can thus be identified: (a) the continental shelf, (b) the continental slope, (c) complex transition zones, (d) the ocean bed or the floor of abyssal basins, (e) the Mid-Oceanic Ridge (Figure 2). These forms of the relief, as will be shown later, are clearly differentiated by the structure of the Earth's crust and by geophysical anomalies, which were undoubtedly caused by global effects.

Within the above-mentioned physiographic provinces there are numerous regional and local forms of submarine relief. They form the observed diversity of the floor structure when one looks at the bathymetric chart or ocean-floor profiles. Their origin is determined by a combined action of endogenous and exogenous factors that form the morphostructures and the morpho*sculptures* on the ocean floor. Everywhere both these groups of factors act jointly and simultaneously; only in some cases the first one prevails and in others, the second. As already noted, in this work the endogenous factors and their effects are primarily dealt with.

To estimate the significance of certain physiographic provinces in the ocean-floor morphostructure, we have measured their areas [59]. The results are shown in Table I. Table II lists the data characterizing the areas, volumes

TABLE I
Areas of the Atlantic Ocean geomorphological provinces

Geomorphological provinces	Area	
	$\times 10^3$ km^2	%
Shelf	7900	9.0
Continental slope	7282	8.3
Continental rise	9345	10.6
Transition zones	5233	6.0
Mid-Atlantic Ridge	24023	27.3
Ocean bed	34102	38.8

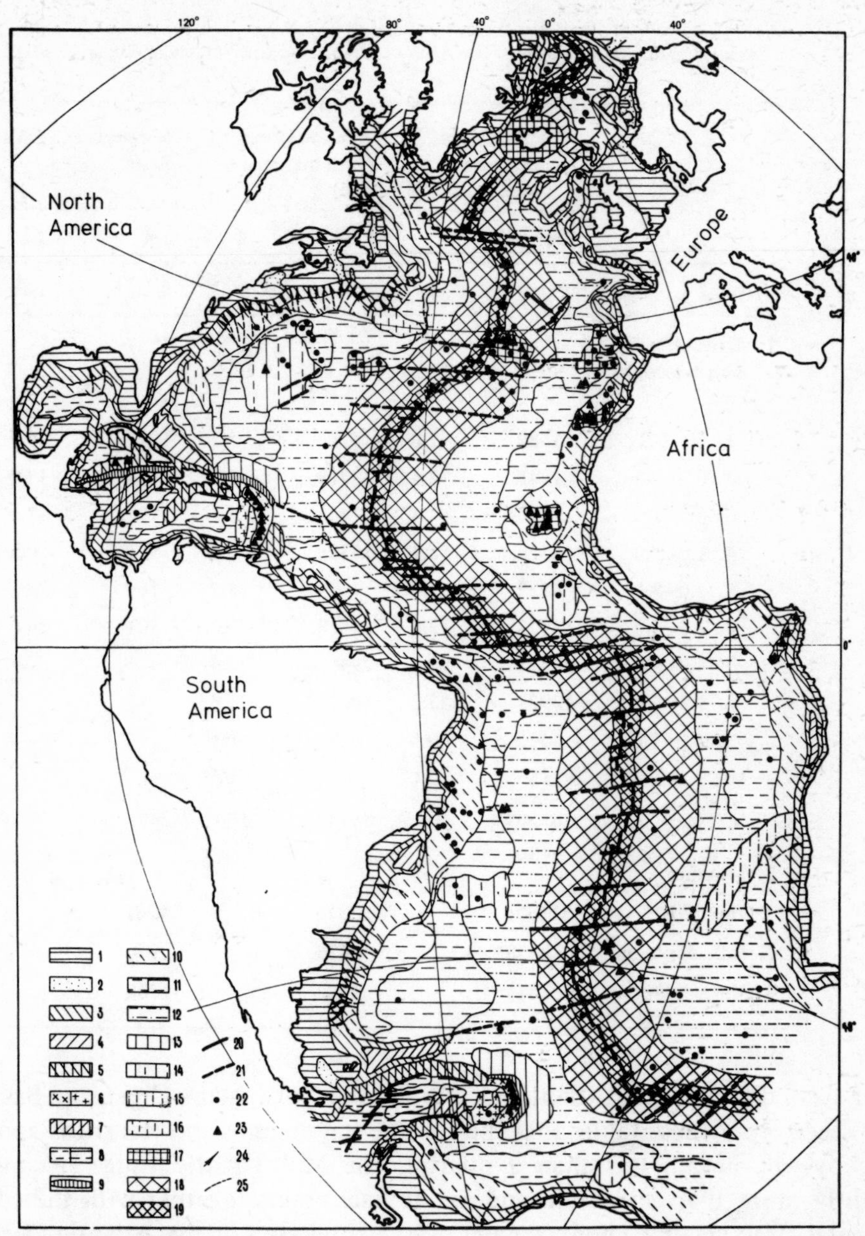

Fig. 2. Geomorphological map of the Atlantic Ocean. (1) abrasive-accumulative shelf plains; (2) shelf basins and trenches; (3) continental slope scarps; (4) marginal plateaus; (5) submarine island arc ridges; (6) modern and ancient volcanic arcs; (7) submarine rises; (8) flat (a) and hilly (b) plains of the floor of marginal seas; (9) deep-water trenches; (10) sloping plains of continental rises; (11) flat (a) and rolling (b) abyssal plains; (12) zones of abyssal hills; (13) marginal outer swells; (14) elevations and plateaus; (15) block ridges; (16) accumulative ridges; (17) volcanic massifs; (18) the Mid-Oceanic Ridge flanks; (19) rift zone; (20) trenches-faults; (21) zones of submarine relief disjunction; (22) volcanic seamounts; (23) volcanic islands; (24) submarine canyons; (25) turbidity flow channels.

TABLE II
Areas, volumes and depths of the Atlantic Ocean and the Norwegian-Greenland Basin

	Atlantic Ocean (without seas)		Atlantic Ocean (with marginal seas)		Norwegian-Greenland Basin	
	$km^2 \times 10^3$	%	$km^2 \times 10^3$	%	$km^2 \times 10^3$	%
1	2	3	4	5	6	7
1. Areas of bathymetric sections, km:						
1. 0–0.2	5689	7.0	7256	8.2	247	9.9
2. 0.2–0.5	1948	2.4	2219	2.5	477	19.4
3. 0.5–1.0	1136	1.4	1443	1.6	173	7.0
4. 1.0–2.0	2626	3.2	3353	3.8	511	20.5
5. 2.0–3.0	6237	7.6	7189	8.2	586	23.5
6. 3.0–4.0	15723	19.4	17618	20.0	501	20.0
7. 4.0–5.0	28473	35.0	29462	33.4	–	–
8. 5.0–6.0	18937	23.3	19005	21.6	–	–
9. 6.0–7.0	512	0.6	542	0.6	–	–
10. > 7.0	155	0.1	57	0.1	–	–
Total	81309	100.0	88144	100.0	2495	100.0
2. Volume, $km^3 \times 10^3$	312576		328315		4261	
3. Max. depth, m	8428		8428		3970	
4. Mean depth, m	3844		3725		1708	

and mean depths of the Atlantic Ocean and the Norwegian–Greenland Basin. It is seen from these tables that the continental margins and transition zones occupy, all in all, less than a quarter, the Mid-Atlantic Ridge occupies slightly more than a quarter, and the abyssal basins together with the continental rises occupy about a half, of the total area of the Atlantic Ocean

There are many islands in the Atlantic Ocean and they can be subdivided into two general types: epicontinental and oceanic. Physiographically and tectonically those of the first type are either directly connected with the adjoining parts of continents or situated on the extensions of continental structures. Among these are such shelf islands as the British Isles, Newfoundland, Greenland, the Falkland Islands, and others, as well as the island arcs:

the Antilles, the South Sandwich and South Orkney Islands and South Georgia. Oceanic islands do not have any direct connection with continents and are exclusively volcanic formations. They are situated either within the Mid-Oceanic Ridge (Iceland, Jan Mayen, the Azores, St Paul Rocks, Ascension, St Helena, Tristan da Cunha, Gough and Bouvet islands) or on the floor of abyssal basins (the Bermudas, Madeira, the Canary Islands, the Cape Verde Islands, and a number of others).

2. CONTINENTAL SHELF

In the structural-tectonic respect, the continental shelf represents a direct extension of the coastal parts of land, mainly of coastal plains. The basement of the shelf is evidently composed of continental structures, with an overlying sedimentary cover of varying thickness almost everywhere, which masks the original bed causing the observed considerable smoothness of the submarine relief. Moreover, the surface of the shelf in a comparatively recent geological past was epicontinental land and has, therefore, passed through the subaerial stage of development. A subsequent rise of the ocean level and the neotectonic subsidence of the continental margins caused the abrasive-accumulative levelling of the shelf surface.

The shelves of glacial regions, such as those in the northern part of the Atlantic Ocean, in the Norwegian-Greenland Basin and in Antarctica, differ most sharply from those in other areas. Their characteristic features are: the dissection of the floor surface by systems of longitudinal and transverse trenches into a number of elevated areas – banks – littoral shoals with a hillocky relief, the widespread occurrence of minor forms of relief on the shelf – hillocks and small ridges on the slopes of banks and trenches (Figure 3). The dissection of glacial shelves is, on the one hand, caused by neotectonic dislocations, as a rule of an inherited character, and on the other hand, by the effect of Quaternary and recent glaciers, that penetrated within the shelf, performed the glacial scouring along the trenches and, after melting, left morainic material on the shelf [25, 53, 163].

With the exception of the zones of coral activity, the rest of the shelf regions may be defined as normal. Their surface has been mainly levelled by abrasive-accumulative processes occurring with changes of sea-level in the Quaternary time [34, 57]. Three principal zones are, as a rule, identified in the relief of the normal shelves: (a) littoral shoal coinciding with the underwater coastal slope and exposed to present-day wave abrasion; (b) median zone with a very even and almost horizontal surface, in some places with sand ridges, submarine valleys and terraces; (c) outer zone with gradually increasing gradients up to the outer edge of the shelf, characterized by submarine terraces. These zones are most distinctly manifested and sufficiently

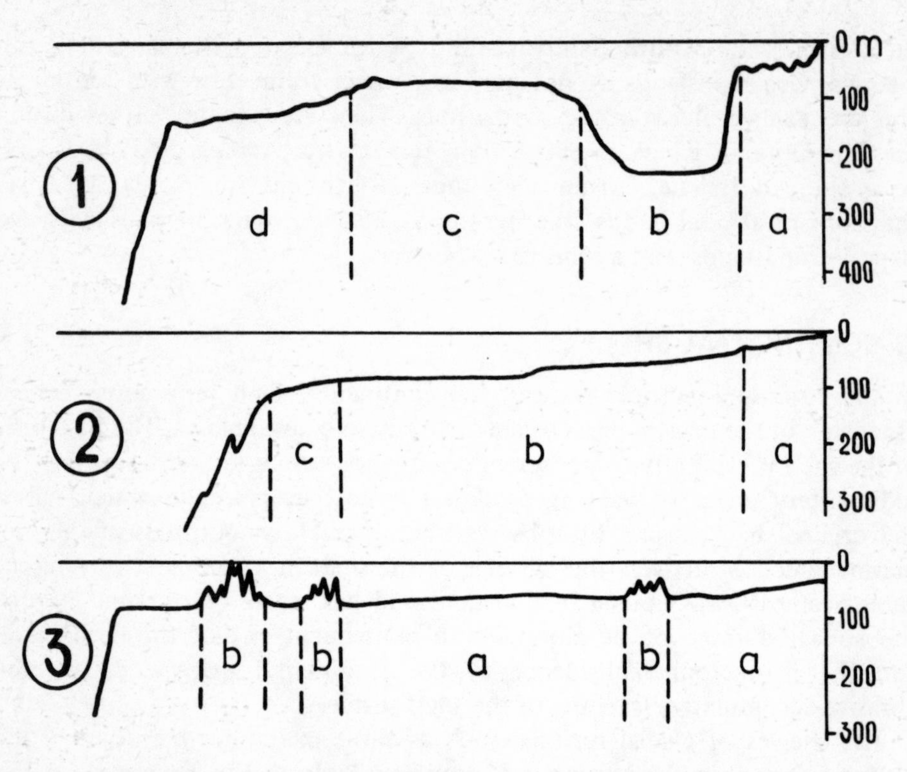

Fig. 3. Typical profiles of the continental shelf relief. (1) glacial shelf: (a) littoral shoal (strandflat), (b) longitudinal trench, (c) the outer shelf bank, (d) outer part of the shelf; (2) normal shelf: (a) coastal zone, (b) median zone, (c) outer zone; (3) shelf with coral structures: (a) shelf surface, (b) coral reefs.

well studied on the shelves of the eastern coast of the U.S.A. and the western coast of Central Africa [81, 223].

Shelf regions with coral structures are mainly constructed in the same way as normal shelves. Their surface, however, is complicated by hills, odd forms of knolls and ridges made by reef-building organisms and their detritus. This type of relief is common in the coastal areas of Central America (the Gulf of Mexico and the Caribbean) and in the north-eastern part of South America.

The Greenland Shelf

The width of the shelf differs along the eastern and the south-eastern coasts of Greenland. In the north it is more than 170 miles wide and in the south it contracts to 30–40 miles. The shelf surface is everywhere dissected by transverse trenches, most of them on the extension of large coastal fiords. Banks of the outer shelf are separated from the littoral shoal by series of longitudinal trenches. Depths in the trenches reach 350–400 m, sometimes more than 500 m, and on the banks are not greater than 200–250 m [54]. A level relief prevails on the banks, whereas a hill-and-ridge relief is predominant on the bank slopes and in the trenches. At the south-eastern coast of Greenland in a southward direction the shelf relief becomes more and more uneven, contrasts of depths and steepness of slopes increase, which seems to be explained by a greater tectonic diversification of the coastal regions of South Greenland [53].

The south-western coast of the island has a similar shelf structure. The width of the shelf here varies from 40 miles in the south to 90 miles near Disco Island. Depths on the banks range from 60 to 150 m, and reach 200–400 m in the trenches. Bank slopes and trenches are complicated by hills and ridges [79].

The North-American Shelf

A typical glacial shelf up to 60–80 miles wide stretches from Baffin Island to Newfoundland. A narrow belt of the littoral shoal with a hillocky relief is separated from the rest of the shelf by a series of steep scarps, sometimes with longitudinal trenches along their basement. Transverse trenches are sufficiently well manifested everywhere, their depth reaching 300–500 m. On the shelf banks the depths are from 150 to 300 m. The top-most surfaces of banks and the trench bottoms are level, and the rest of the shelf territory is characterized by a relict glacial-accumulative hill-and-ridge relief [79, 163].

The shelf width increases to 250 miles south-east of Newfoundland, where the Grand Newfoundland Banks are situated with depths of from 50 to 100 m. Further to the south-west, along the coast of Nova Scotia, there is a shelf up to 100–120 miles wide, divided into three parts by two large transverse trenches, with depths of 400–450 m and 250–300 m, respectively, coming out from the Gulf of St. Lawrence and the Bay of Fundy. In turn, these are divided by small trenches into a number of banks with depths over them not exceeding 100 m [79]. The surface of these banks is, for the most part, levelled and in places dissected by sand ridges formed by tidal currents. This is most clearly seen on the Georges Bank to the south of the Bay of Fundy. Recent geological and geophysical investigations have shown that the banks of the Nova Scotia shelf, and evidently, of other glacial shelves, are made up of Neogene sediments, overlain by a thick moraine layer [233].

To the south of Georges Bank the appearance of the shelf changes radically. The glacial-accumulative forms disappear. The shelf surface up to the southern extremity of Florida is mainly level and depths are usually less than 100 m. The width of the shelf decreases from 70 miles in the north to 10 miles in the south but amounting only to 2 miles in the Miami area. The outer edge of the shelf to the north of Cape Hatteras has depths of up to 100 m, and to the south they decrease to between

50 and 60 m [233]. The coastal zone down to depths of 20–30 m is complicated by small sand ridges orientated, as a rule, along the coastal line. The main part of the shelf is almost horizontal, has depths equal to 40–50 m and, in places, is incised by submarine valleys. Very well defined is the valley situated on the prolongation of the Hudson River. It cuts into the surface of the shelf to depths of 20–30 m. The valley reaches the outer edge of the shelf and is continued along the continetal slope as the Large Hudson Canyon.

Also encountered on the surface of the shelf, and elongated over a considerable distance, are low banks with adjoining flat portions of floor, which are relicts of submerged coastal terraces. There are several levels of these terraces at the depths of 35–40, 55–60, 75–80, 110–120, and 150–160 m [151]. Similar terraces are also known to exist in other regions of the continental margins of the Atlantic Ocean, which is indicative of global changes of the ocean level during the Pleistocene. The lower of the terraces (depths 150–160 m) seem to have been formed when the ocean level went down during the maximum glaciation, and the terraces lying above were formed during the latest glaciation and its subsequent rise in level.

The Mexicano-Caribbean Shelf

A rather wide shelf (up to 100–130 miles) stretches along the south coast of the U.S.A. in the Gulf of Mexico. Its surface is, for the most part, levelled with the exception of coral reefs scattered in some places, Along the east coast of Mexico the shelf width decreases to 10–20 miles and in the southern part of the Bahia de Campeche it is less than 5 miles. At the western and northern shores of the Yucatan Peninsula, however, the shelf widens again up to 100 miles, forming the extensive Campeche Bank with depths over it equal to 50–80 m. All along the shelf is a gently sloping plain diversified by numerous coral formations appearing as ridges and hills of irregular shape [11].

At the eastern coast of Yucatan, the shelf width again decreases to 2–5 miles forming a strip of uneven relief with numerous submarine and above-water coral reefs. But to the north-east of Honduras and Nicaragua the shelf forms a prominence up to 120 miles wide – the Mosquitos Bank. The depths over the bank are up to 60–80 m. In its coastal part numerous coral reefs are encountered.

From the Mosquitos Bank right up to Trinidad the shelf of the Caribbean Sea is a narrow strip (10–20 miles wide) of uneven relief with numerous coral reefs and tectonic banks. At the Venezuela coast between Caracas and Isla de Margarita the shelf is incised by the tectonic Cariaco Trench with depths reaching 1420 m. It is separated from the continental slope by a narrow rise with the depths over it not exceeding 130 m.

The South American Shelf

To the south-east of Trinidad the shelf widens to 50–70 miles and near the mouths of the Amazon it reaches a width of 170 miles, while depths at its outer edge are equal to 70–90 m. The shelf surface is mainly level, but in the coastal zone is diversified by coral reefs with depths over them of less than 20 m [34]. In places relicts of coastal terraces can be traced at depths of 30–50 and 70–80 m. On the extension of rivers poorly expressed submarine valleys are seen, the largest being situated on the extensions of the Amazon and the Rio Tocantins.

Along the east coast of Brazil the shelf is narrow (15–20 miles wide), but between latitudes 16° and 20° South forms two prominences – the Royal Charlotte and the Abrolhos Banks. The shelf surface is everywhere a gently sloping plain, diversified in the coastal zone by numerous coral reefs and in the outer zone by sparse knolls that seem to be subsided ancient coral structures [34]. The depths at the outer edge of the shelf are equal to 50–70 m.

To the south of the Abrolhos Bank the shelf width gradually increases. Near the Rio dè la Plata it is equal to 100 miles and at the Falkland Islands 400 miles. The depth at the outer edge of the shelf also gradually increases from 70 to 180 m. Coral constructions on the shelf disappear and its surface becomes a gently sloping plain with well-defined littoral, median and outer zones [186]. In the bay of Rio de la Plata and in the Bahia Blanca, on the even floor, ill-defined submarine valleys can be traced. In the southern part of the region, near the coast of Patagonia, and also around the Falkland Islands, at the depths of about 50 m lies a zone of uneven hillocky relief, indicating a considerable role of glacial-erosion processes that took place during the Pleistocene glaciations.

The West-European Shelf

The northern part of the region, situated within the Norwegian-Greenland Basin is a typical glacial shelf. Along the west coast of Spitsbergen extends a relatively narrow shelf (up to 40 miles wide), incised into a series of banks by several transverse trenches. Depths in the trenches are equal to 250–350 m, while on the banks they are less than 200 m. The surfaces of banks and trench bottoms are level, and a hillocky relief is characteristic of the rest of the shelf territory.

Between Spitsbergen and Scandinavia the boundary of the Norwegian–Greenland Basin passes along the western outskirts of the Barents Sea, which is one of the largest shelf areas on the Earth. Large trenches reach the edge of the shelf – the Südkap and the Vest Troughs (with depths

370–390 and 450–480 m, respectively) – and the Spitsbergen Bank situated between them, with the small Bear Island on it, are the main forms of relief here [54].

The shelf along the coast of Norway can be divided into three parts. At the north-western and south-western coasts the shelf is narrow (20–40 miles), the depths on banks are between 60 and 150 m, in the trenches 300–400 m. In the Central Norway region the shelf widens to 120–140 miles, trenches and banks become more extensive. Depths in the trenches reach 350–450 m, and on the banks are from 170 to 280 m. Along the coast extends the zone of uneven, hillocky relief with depths up to 40–50 m, called a strandflat, similar to the coastal zones of other glacial shelves. Its relief has been mainly formed under the effect of glacial erosion and wave abrasion.

The floor of the trenches on the Norway shelf has been levelled as a result of sedimentation, and smoothness of the surface of banks with depths less than 100 m has been caused by the action of abrasive-accumulative processes during the post–glacial rise of ocean level. The hilly relief on the rest of the shelf territory indicates the presence of relict glacial-accumulative forms created in the Pleistocene, when glaciers covered the shelf, and the ocean level was getting lower [81, 163].

An extensive shelf is located under the waters of the North Sea. In its north-eastern part, skirting the Norway coast, stretches a broad trough-like trench with depths equal to 300–400 m (in the Skagerrak – more than 600 m). The rest of the shelf territory is a plain slightly inclined to the North, with depths gradually increasing from 30–40 to 150–180 m. Numerous small banks are scattered on it. These are the relicts of terminal morainic ridges left by the Scandinavian Glacier. There are also occasional narrow linear depressions more than 200 m deep, which may be the relicts of ancient river valleys crossing the sea floor when it was still land.

Along the western coasts of Scotland and Ireland the shelf width varies from 20 to 70 miles, which is caused by the complex outline of the shore line. All the way from the Shetland Islands to the English Channel the shelf surface is a wavy plain with individual banks (depths less than 100 m) and depressions (depths more than 200 m). The outer edge of the shelf has depths equal to 150–170 m. To the south-west of Ireland poorly defined terraces are noted at depths of 30–40 and 60–80 m, and in the outer part of the shelf small banks in the form of ridges made up of moraine material are found. To the south of the English Channel glacial-accumulative forms on the shelf disappear [34].

Along the western coast of France the shelf width gradually decreases southward from 80 to 25 miles. Its surface is for the most part even [115]. The outer edge has depths ranging from 150 m in the north to 100 m in the south. In places, at depths of about 70–80 m, terraces are observed. Submarine valleys extend from the mouths of large rivers, such as the Loire and the Garonne, but these do not reach the edge of the shelf.

Along the coast of the Iberian Peninsula the shelf is narrow (10–20 miles) and has a block structure caused by tectonic movements of the continental margin. Shallow areas alternate with deeper ones (depths up to 200–250 m). The hillocky relief is characteristic for the coastal zone, while the relief of the rest of the shelf territory is predominantly even.

The African Shelf

From the Strait of Gibraltar to the area of the Canary Islands the shelf width does not exceed 25 miles and at its outer edge reaches 150–190 m. The shelf surface is mainly even, but opposite the promontories it is diversified by mounds and ridges of structural origin. Further to the south the depth of the outer edge of the shelf gradually decreases, amounting to no more than 100–110 m near the Cabo Juby. Further on, these depths remain almost unchanged to the Gulf of Guinea, only occasionally decreasing to 80–90 m or increasing to 120–130 m. The shelf width also varies. Opposite the promontories it is reduced, amounting near Cape Verde, for example, to no more than 3–4 miles. To the south-east, in the vicinity of Conakry and Freetown the shelf width increases to 100–110 miles, and from there on is equal to 20–40 miles. Against the background of an even relief, here and there, especially near the prominences, are seen small hills and ridges of structural origin. On the prolongations of large rivers, such as the Nunjes, the Konkoure and others, extend submarine valleys, incising, as a rule, the whole width of the shelf [194]. Opposite Abidjan on the prolongation of the ancient channel of the Comoe River there is a large submarine valley, Trou Sans Fond, in the form of a trench with a maximum depth of 400 m.

From the Bay of Biafra to the mouth of the Congo the shelf width does not exceed 35 miles, and the outer edge is situated at the depths of 100–110 m. The shelf surface is even, with the exception of the area near the Cape Lopez, where mounds and ridges are encountered. From the mouth of the Congo, across the whole shelf, extends an almost rectilinear trench with depths up to 800 m; it then continues on the continental slope in the form of the huge Congo Canyon.

To the south of the mouth of the Congo the shelf width gradually decreases and the depth at the outer edge increases. Between the latitudes 16° and 20° South the beach almost immediately passes into the continental slope without any distinct discontinuity in the floor profile. Further to the south, however, the shelf begins to expand and near Walvis Bay reaches a width of 70 miles. The depths at the outer edge of the shelf are 100–110 m.

The shelf surface is mostly even, and only in the vicinity of latitude 20° South are there outcrops of bedrocks in the form of small ridges [60].

To the south of Walvis Bay the shelf in two places (at latitude 26° South and near Capetown) narrows to 10 miles, and the rest of it is 50–70 miles wide. Depths at the outer edge are 230–250 m, near Capetown – even up to 400 m [191]. But on the Agulhas Bank, situated near the southern extremity of Africa, depths do not exceed 180 m. All along its length the shelf surface is mainly even, and (in places) diversified by gently sloping ridges.

The Antarctic Shelf

Along the southern boundary of the Atlantic Ocean, near the shores of Antarctica, stretches a band of the shelf with characteristic properties of continental glaciation regions [26]. In the western part of the region, in the Weddell Sea, the shelf has a considerable width, whereas to the east, along the coast of Queen Maud Land, it is narrow. All the coastal part of the shelf is covered with glaciers. The outer edge of the shelf in the Weddell Sea is located at depths of 300–350 m. In the south-eastern part of the sea there is traced a large trench on the shelf, whose depths are equal to 1400–1500 m. From the outer edge of the shelf it extends southward below the Filchner Ice Shelf.

Along Queen Maud Land individual glacier tongues reach the edge of the shelf. The structure of the subglacial part of the shelf is little known, but, by analogy with other investigated areas, one can assume the presence here of longitudinal and transverse trenches. Here and there on the open parts of the shelf appear the mouths of transverse trenches, where the depths are greater than 400 m. On the remaining part of the shelf the depths are equal to 200–300 m. The shelf surface is mainly hillocky, near the bedrock coast it is moundy with distinctive forms of dissections.

3. CONTINENTAL SLOPE

The zone of transition from the continent to the ocean bed has a different structure, determined by the structural features of the Earth's crust and the history of the geological development of continental margins. In the Atlantic Ocean a simple type of transition prevails – a continental slope that is morphologically represented by an enormous ledge here and there diversified by banks, block forms of dissection, submarine canyons. The gradients of continental slope surface differ from 1–2° to 10–15°, and on some banks they are considerably higher. On the steep parts of the slope sedimentary or metamorphosed bedrocks are, as a rule, exposed, while the more gently sloping parts, particularly the lower parts of the slope, are covered with sediments. By their genesis the continental slope banks can represent either faults (or series of faults), or flexures of the continental crust towards the ocean bed, or the outer slopes of accumulative terraces denudated by exogenous processes on continental margins and made up of sequences of sediments and sedimentary rocks.

At the foot of a continental slope thick accumulative series are usually formed, extending in the form of inclined plains into the floor of the ocean. Morphologically these series represent the lowest parts of the slope, and so they are frequently regarded as transition zones between continents and ocean floor and are named continental rises [46, 151]. Structurally, however, the series represent accumulative formations on the marginal parts of the ocean bed, while the boundary between the floor of the ocean and the continental slope is buried under sediments and, strictly speaking, located somewhat beyond the presently observed foot of this slope [57].

Large bulges on the continental slope, reaching tens and hundreds of miles in width, form the marginal plateaus. As indicated by geophysical data they are blocks of continental margins subsided to a different depth and mostly overlain by a sedimentary cover. Small banks, up to several miles in width, are represented either by ancient coastal terraces (usually to depths of no more than 500–800 m) or structural and tectonic forms [151].

Submarine canyons incising the continental slope are valleys, with a narrow bottom and steep side slopes, deeply entrenched into their surface. Most of them are located only within the continental slope but the largest are the continuations of shelf submarine valleys at river mouths. It is generally thought that the canyons are primarily of tectonic origin with subsequent reworking by the action of river flows in subaereal conditions during the lowering of ocean level (for the upper parts of canyons) and by the effect of underwater turbidity currents [35]. Moreover, turbidity currents are the agents of transporting terrigenous material carried off from the shelf along canyons onto the ocean floor. Among other things, this is indicated by many of the canyons having extensions in the form of channels within the sloping plains of accumulative series.

The Continental Slope of Greenland

Along the east coast of Greenland the continental slope has the form of a ledge with average inclination of 3–4° and a slight block dissection [54]. The foot of the slope is situated at a depth of 2800–3200 m where it gradually passes into an inclined plain of the accumulative series. To the south of latitude 72° North along the foot of the slope lies the Greenland-Iceland Trench 1500–1600 m deep, which in the Denmark Strait becomes shallower and separates the projections of the Greenland and the Iceland shelves forming the Greenland-Iceland Rise. The depth of the trench here is 590–600 m.

To the south-west of the rise the continental slope structure is also rather simple. Its average gradient is 3–5°. The foot goes down to depths of 2100–2200 m where it smoothly passes into the inclined plain of the Irminger Basin. As one approaches the southern extremity of Greenland, however, the slope angle increases to 8–10° and in some places exceeds 15°. The slope here has a characteristic valley-and-block dissection and its lower portion is complicated by asymmetric uplifts of fault origin [53]. The depths at the foot of the continental slope near the southern extremity of Greenland reach 2800–3000 m (Figure 4).

The continental slope at the south-western coast of Greenland is of a similar structure with characteristic features of valley-and-block dissection [67]. The gradient of its upper part reaches 10–15°, while the lower part slopes more gently and is diversified by asymmetric blocks. The depths at the foot of the slope gradually diminish northwards as one approaches the Baffin-Greenland Rise. The surface of the latter is slightly inclined to the east and has depths from 400 to 800 m. The southern slope of the rise is gentle and gradually passes into the floor of the Labrador Basin.

The Continental Slope of North America

From the Baffin-Greenland Rise all the way up to Flemish Cap lies a rather monotonous continental slope, mostly gentle and only slightly dissected. The inclination of its upper part is 2–3°. There are small canyons and terraces here and there. The lower part of the slope at depths of about 2500 m smoothly passes into the plain of accumulative series.

To the east of the Grand Newfoundland Banks, separated from them by a trench up to 1200 m deep, is located the rounded Flemish Cap bank representing a highly elevated marginal plateau. Depths over the bank range from 150 m to 350 m, and the slopes have a convex profile with a steepness of up to 15–18° [79].

From Flemish Cap to Cape Hatteras the continental slope forms a concave bank, strongly dissected by numerous submarine canyons and structural terraces. The gradient is almost constant throughout and amounts on the average to 4–6°, but at some points it reaches more than 10°. The foot of the slope is situated at depths of from 2800 to 3600 m. A broad inclined plain of accumulative series is located below, subdivided into the upper and the lower parts [151, 223]. The largest submarine canyons, such as the Hudson, the Oceanographer, the Hully and others, continue in the form of suspension flow channels far into the accumulative series.

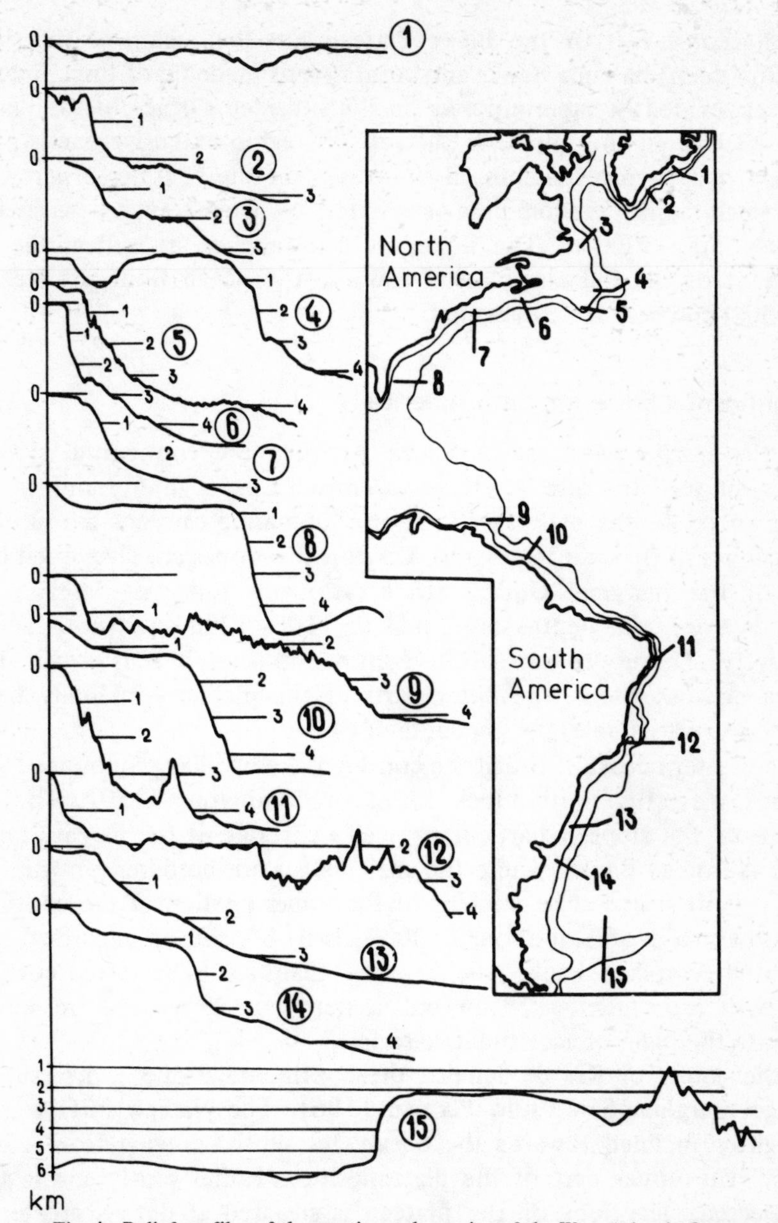

Fig. 4. Relief profiles of the continental margins of the West Atlantic Ocean.

To the south of Cape Hatteras the continental slope is diversified by the large marginal Blake Plateau, whose width reaches more than 150 miles. Its surface is mainly even, slightly inclined to the east and has depths of from 900 to 1200 m. The outer slope of the plateau is very steep (20–30°) and represents a typical example of a fault zone on the continental slope. Its foot goes down to a depth of 4500–5000 m.

To the south-east of the Blake Plateau lies the region of the shallow (10–20 m deep) Bahama Banks and coral islands made up of thick limestone sequences, evidently superimposed on the subsided surface of the marginal plateau [197]. In the northern part of the region extensive banks prevail, separated by narrow trenches. By contrast, in the south-east part of the region small banks predominate, separated by wide trenches with depths of up to 1500–3000 m. The slopes of the trenches, as well as the outer escarpment of the Bahama Plateau towards the ocean floor, are steep (up to 20–30°) and very little dissected.

The Continental Slope of South America

Along the north-east coast of South America the continental slope has a rather simple structure [34]. Its average slope is 4–6°, and the foot goes down to depths of 3300–3500 m. Submarine canyons are rare. Near Paramaribo and the mouths of the Amazon the slopes are diversified by the ledges of the marginal Guiana (Demerara) and Amazon Plateaus. Their surface is even, the depths are equal to 1100–1200 and 1400–1600 m, respectively. The upper part of the continental slope here is gently sloping and has no dissections, the outer parts of the plateaus gradually become steeper downwards, reaching gradients of 6–8°.

At the eastern coast of Brazil the continental slope has pronounced valley-and-block dissection with block forms predominating [150]. The block structure of the slope is particularly well expressed at the Royal Charlotte and the Abrolhos Banks, where the shelf ledges are bordered on three sides by steep fault scarps. The gradient of the upper portion of the continental slope (down to depths of about 2500 m) is 6–8°, and on individual scarps, e.g., at the Abrolhos Bank, reaches more than 15°. The lower portion of the slope is more gentle (2–3°) and at depth of 3500–3800 m gradually passes into the plain of accumulative series.

To the south of Rio de Janeiro the continental slope is diversified by the large marginal São Paulo Plateau [186]. The plateau surface is even and slightly inclined towards the ocean, its depths ranging from 2200 to 2700 m. The upper part of the plateau slopes rather gently and is almost non-dissected. The foot of the plateau is situated at depths greater than 3500 m.

Along the shores of Argentina the continental slope has an average gradient of 3–5° and is dissected by numerous submarine canyons. Its foot bends smoothly down and passes into the plain of the accumulative series, the depth of the foot gradually increasing to the south from 3500 to 5000 m.

To the east of the Falkland Islands is located the marginal plateau of the same name, stretching over more than 600 miles [3, 186]. The plateau surface is slightly inclined to the east and in the middle part has depths of 2500–2700 m. The eastern part of the plateau is elevated in the form of a large block, with the depths over it equal to 1300–1500 m. The northern and southern slopes of the plateau are formed by almost rectilinear ridges with inclinations equal to 10–12° and 5–6°, respectively.

The Continental Slope of Western Europe

Along the Western Spitsbergen and in the Spitsbergen Bank region the continental slope is rather steep (4–6°) and has a slight block dissection. Opposite the Südkap and the Vest Troughs the slope is, however, very gentle (about 1–2°) and formed by vast fans, probably made up of sediment sequences, brought from the Barents Sea shelf over a long period of time [54].

At the north-western coast of Norway the continental slope is steeper (on the average 8–10°, sometimes 20°) and dissected by numerous canyons of tectonic origin [51]. The foot of the continental slope here reaches depths of 2700–2900 m.

In the Central Norway region the upper part of the continental slope goes down gently and at depths of 1300–1400 m smoothly passes into the surface of the marginal Norwegian Plateau (or the Voring Plateau). Its relief is gently undulating and even (Figure 5). The outer slopes have gradients reaching 6–7°, and their feet are situated at depths of 3100–3500 m.

The region to the north-west of Great Britain is a characteristic area of complicated block structure. It includes the Faeroe Islands with the surrounding shelf, the submarine Wiville-Thomson Ridge and the Iceland-Faeroes Rise, the Rockall Plateau, the Ireland and the Faeroes-Shetland Trenches, and also, probably, the Iceland Plateau. This region was probably formed as a result of uneven subsidence of ancient continental blocks along fault zones, mostly of north-eastern strike, during the Tertiary [55].

The shelf around the Faeroes lies mainly at the depth of up to 170–190 m (in individual depressions – up to 350 m). From the south and south-east the shelf is skirted by the wide Faeroes-Shetland Trench, where the depths gradually increase from 1100 to 1700 m in the direction from the south to the north-east.

To the south-west of the Faeroes lies the vast Rockall Plateau, separated from the continental shelf by the wide Ireland Trough [30]. The bottom of the trough is even and inclined to the south, so that the depths gradually increase from 1600 to 3100 m. It is separated from the Faeroes-Shetland Trench by the Wiville-Thomson Ridge, a narrow swell (not more than 30 miles wide) with depths ranging from 380 to 620 m. The Rockall Plateau in its northern part is formed by a broad swell dissected by transverse depressions into a series of rounded rises (the Faeroes, Bill Baileys, Outer

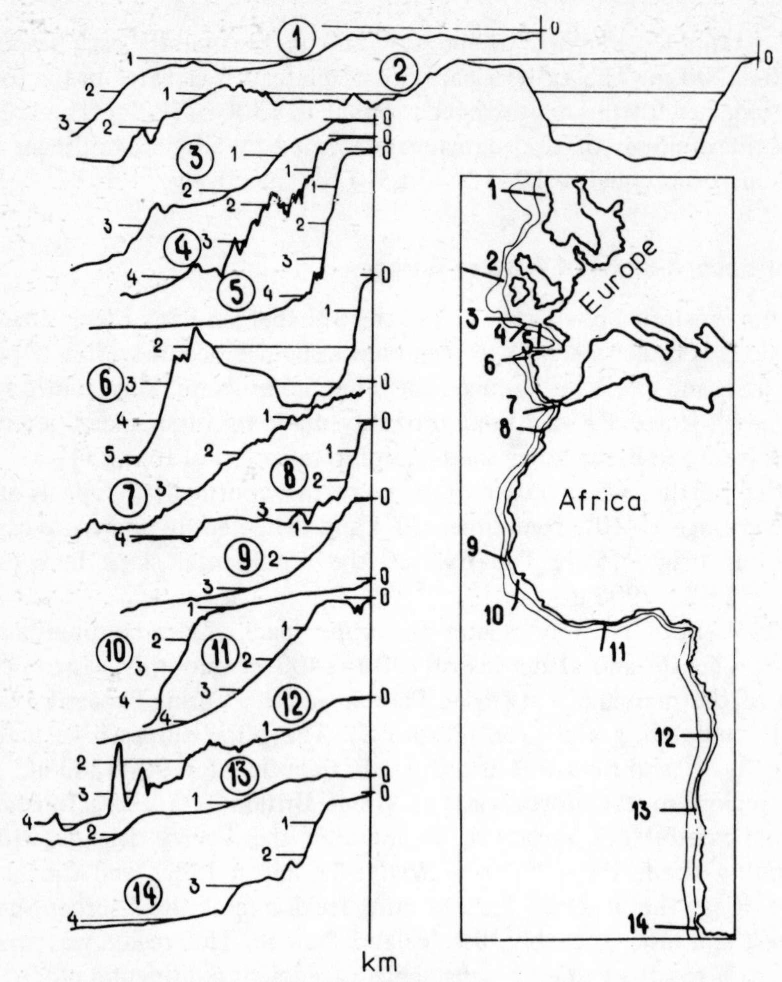

Fig. 5. Relief profiles of the continental margins of the East Atlantic Ocean.

Bailey, and George Bligh Banks), the depths over them gradually increase southward from 80–90 to 400–500 m. The southern part of the plateau is more extensive and divided by a longitudinal depression with a depth of up to 1200 m into two gently sloping crests: the western – Hutton Bank and the eastern – the Rockall Oceanic Bank. The depths over the Hutton Bank amount to 600–700 m, while those on the Rockall Oceanic Bank are not more than 180–190 m. The outer slopes of the Rockall Plateau have a gradient of 3–5° and are diversified by large block dissections.

Between the Faeroes and Iceland stretches the wide Faeroes-Iceland Rise [51]. Its surface lies at depths ranging from 300 to 450 m. In the central part of the rise there are small elevations, bounded by steep scarps 40–80 m high, probably the ancient shore cliffs. The outer slopes of the rise are gentle but their gradient increases considerably with depth.

The Iceland Plateau, located between Iceland and Jan Mayen, is a rolling plain with depths of 1800–2200 m, diversified by rather rare hills and scarps orientated in the north-east direction. On the eastern side of the plateau, to the south of Jan Mayen, stretches the block Jan Mayen Ridge, with depths over it ranging from 600 to 900 m. Its southern extremity is broken up into a number of asymmetric blocks. According to the results of investigations made on board the R/V *Akademik Kurchatov* the Iceland Plateau is a submerged portion of the continental crust of the marginal plateau type, that was transformed to some extent by riftogenic processes during the formation of the Mid-Oceanic Ridge [88]. To the west of Ireland in the continental slope zone is the Porcupine Bank, separated from a shelf by a shallow trench. Its depth is equal to 200–350 m, and it may be regarded as a not deeply submerged marginal plateau. The outer slopes of the bank are steep (5–6°), and their foot reaches the depths of 4000–4200 m.

To the south of the Porcupine Bank up to the south-eastern corner of the Bay of Biscay the continental slope is represented by a rather steep scarp (inclinations of 6–8°, in places up to 10°), dissected by numerous submarine canyons [115]. The foot of the slope is situated at a depth of between 4500 and 4600 m, where it borders on a narrow strip of accumulative series. The depths of entrenchment of the canyons into the surface of the slope reach 300–500 m, the slope of their walls is equal to 10–15°. The largest canyons – the Cap Ferre and Cap Breton Canyons are situated opposite the mouths of the Garonne and the Adour rivers.

The continental slope along the northern and western coasts of the Iberian Peninsula has a complicated block-fault structure. Its gradient varies within a fairly broad range, on ridges reaching more than 20°. The foot of the slope is situated at the northern coast at depths of 4600–4700 m, at the western coast – at depths of about 5000 m. Near the north-western coast the continental slope is diversified by the marginal Iberian Plateau with a hilly surface and depths of 1500–1700 m. Small seamounts, the Galicia, the Vigo, and the Porto, are located on it with minimum depths over them equal to 500–600 m [109].

The Continental Slope of Africa

From the Strait of Gibraltar to the Canary Islands the continental slope has an inclination equal to 4–6° and is diversified by occasional canyons and terraces. The foot of the slope reaches depths of 2500–2800 m, where it passes into the surface of accumulative series. The Canary Islands archipelago is separated from the scarp of the continental slope by a trench with a depth of up to 2000 m. Between the Canary and the Cape Verde Islands the continental slope has a mean gradient of 3–5°. Steeper parts are dissected by submarine canyons, the largest of them – the Cayar Canyon – is situated to the north of the Cape Verde Islands, the gentle parts of slope being practically undissected. The foot of the slope is situated at the depths of 2600–2800 m, where it gradually passes into an inclined plain of the accumulative series.

To the south of the Cape Verde Islands up to the mouths of the Niger River the continental slope is steeper (8–10°) and has a stepwise block structure. There are submarine canyons, the largest of them, the Trou Sans Fond Canyon, in this area. The foot of the slope goes down to depths of 3500–3800 m, its deepest part adjoining the Guinea Basin. To the west of Conakry the continental slope is diversified by the Guinea Marginal Plateau. Its surface is slightly inclined towards the ocean, the depths range from 600 to 1000 m. The steep southern scarp of the plateau stretches along the Guinea Fracture [176].

In the vicinity of the mouths of the River Niger the continental slope, where intensive sedimentation has occurred, forms a massive fan, extending towards the ocean [189], with a gently sloping (inclinations about 2°) and levelled surface. The foot is located at a depth of more than 3500 m where it smoothly passes into the plain of the accumulative series.

From the Bay of Biafra to the mouth of the River Congo extends a continental slope of a rather simple structure . Its gradient is small (2–3°), and the foot reaches a depth of 2500–3000 m. The submarine Congo Canyon incising the continental slope in its upper part has a depth of entrenchment of up to 1000 m, the slope of its sides exceeding 20°. From the canyon on the surface of the plain of the accumulative series, suspension current channels extend in a fan-like fashion [214].

To the south of the Congo up to latitude 14° South the continental slope is complicated by the Angola Marginal Plateau, its outer edge reaching depths of 2000–2500 m. The plateau has a steep stepwise slope (inclinations up to 20°) and passes into the plain of accumulative train.

Further up to the Agulhas Bank the continental slope is a scarp with mean gradients equal to 3–5°, diversified here and there by slightly pronounced forms of block dissection and occasional submarine canyons. The foot of the slope reaches depths of about 3500 m, only near the Agulhas Bank increasing to 5000 m. In the region of latitude 20° South the continental slope at depths equal to 1300–1400 m is complicated by a large terrace, bounded from the outer side by a scarp with a gradient of up to 3°. Its foot is located at the depth of 2300 m where it adjoins the block Walvis Ridge [60].

The Continental Slope of Antarctica

This has predominantly a block structure. The simplest structure is that of the slope in the Weddell Sea, where it is represented by a ledge with an inclination of 3–5°. Its foot is situated at depths of 3100–3400 m, where it gradually passes into a well-pronounced plain of the accumulative series. Along the coast of Queen Maud Land numerous forms of block fault dissection result in the angular shape of the continental slope. On rectilinear parts the inclination of the slope varies from 3 to 8°. On the parts of intensive dissection, particularly on the lateral slopes of projecting blocks, inclinations reach 10–15°. The foot of the slope here is situated at depths of 3800–4300 m.

4. TRANSITION ZONES

Transition zones of a complex type are located, as already mentioned, in two regions of the Atlantic Ocean: the Mexicano-Caribbean and the South Antillean. They constitute provinces of Meso-Cenozoic continental structures and their submarine extensions, combined with suboceanic basins of marginal seas and deep-sea trenches. Here complex processes of the Earth's crust transformation and active differentiated tectonic movements are taking place, and this is reflected in sharp contrasts of submarine relief, seismicity and volcanism [6, 97].

In transition zones the ridges of island arcs separate the basins of marginal seas from the ocean bed. On the side of the steepest bend of these arcs extend deep-sea trenches reaching depths of more than 8000 m. The basins of marginal seas are usually separated by submarine rises or ridges that are the branches of island arc ridges. Basin floors are the regions of intensive sedimentation, with their own level of accretion topography depending on the structural status and the history of development.

The Mexicano-Caribbean Zone

The simplest structure is that of the Gulf of Mexico Basin [133]. Here the continental slope is formed by steep step-like zones to the west of Florida (the Florida Fault) and around the Campeche Bank, where its inclination reaches 20–25°. To the south of the mouth of the Mississippi the slope is formed by a massive alluvial cone with suspension current channels traced on its gently sloping surface. In the southern part of the Gulf are encountered the numerous small rounded Sigsbee Knolls, the tops of salt domes covered with sediments. The flat floor of the Gulf of Mexico Basin, forming the Sigsbee Abyssal Plain, is 3750–3770 m deep.

The structure of the floor of the Caribbean Sea is much more complicated [2,11,133]. From the north and the west it is embraced by the Antillean Ridge arc, with the Greater Antilles (Cuba, Hispaniola, Puerto Rico) and the Lesser Antilles (the Leeward and Windward Islands) rising on it. The south-eastern branch of the Lesser Antilles is located on a submarine ridge extending along the coast of Venezuela. The straits between the islands are formed along obliquely transverse faults, dissecting the whole Antillean Ridge into a number of blocks.

The eastern (bow-shaped) part of the Antillean Ridge is formed by a massive swell with a convex top surface, the depths over it being equal to 600–700 m, and numerous volcanic islands are superimposed on it. In the northern part of this arc, on the outer side, extends a second arc of small limestone islands, superimposed on volcanic basement. The side slopes have a step-and-block structure indicative of fault dislocations.

The basins of the Caribbean Sea are characterized by a diversity of areas and depths (Figure 6). The Yucatan Basin is situated between Cuba and Yucatan. The continental slope, bordering it on the west, is formed by a steep stepwise scarp. On the south-east, along the Cayman Ridge, extends a belt of hill-and-block relief. The rest of the basin is occupied by an abyssal plain with depths of 4500–4700 m.

The submarine Cayman Ridge is on the extension of the Sierra Maestra

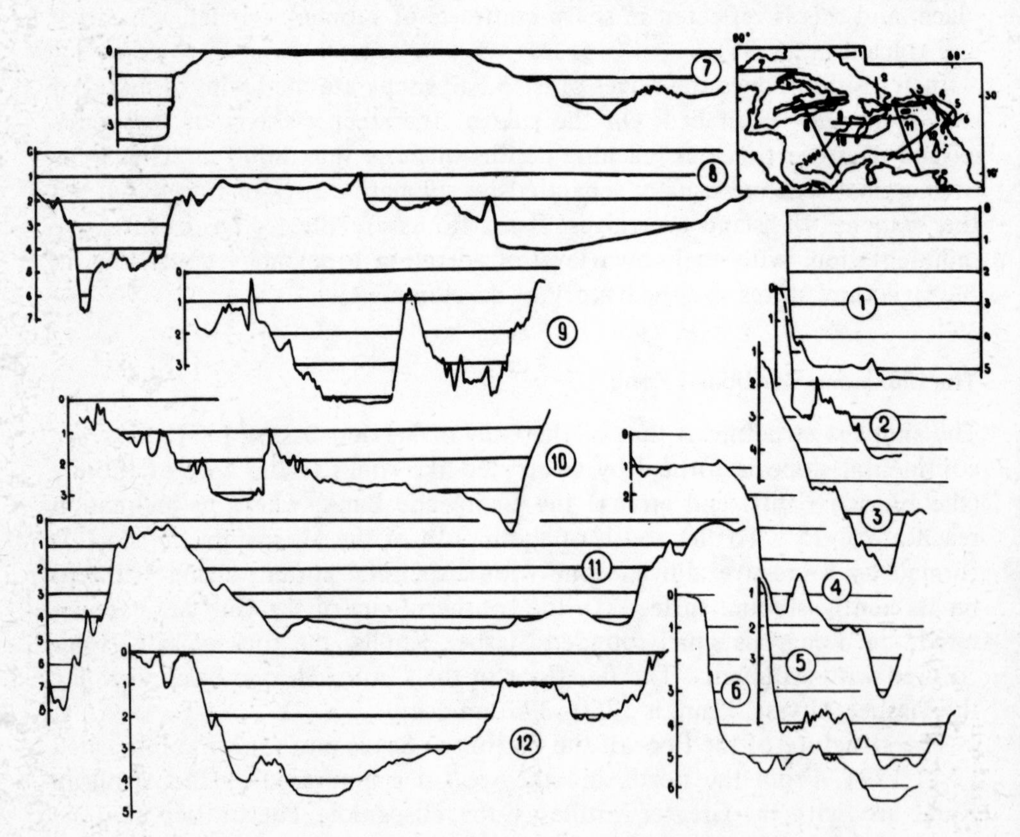

Fig. 6. Relief profiles of the Mexicano-Caribbean Transition Zone.

(Cuba) mountain chains and seems to be structurally connected with them. This ridge represents a massive swell with a step-like top surface and steep slopes. On its elevated steps are situated the Grand and Little Cayman Islands and the Misteriosa Bank. Along the southern slope of the ridge extends the deep-sea Cayman Trench, its maximum depth in the Bartlett Deep reaching 7065 m [65]. The trench has steep stepwise slopes and a narrow floor formed by a chain of elongated depressions separated by gentle rises.

The submarine Nicaragua Rise stretches from the Mosquitos Bank to Hispaniola including also the Rosalind and the Pedro Banks and Jamaica. Its upper surface has a stepwise structure. The saddles between elevated parts (banks) lie at depths of more than 500 m, and the maximum depth of the rise is equal to about 1300 m [1]. A stepwise structure is also characteristic for the side slopes of the rise.

The Colombian Basin stretches from the Panama coast to Hispaniola. Its bottom is slightly inclined to the north-east, the depths gradually increase from 3300 to 4200 m. The surface of the floor is level, indicating prolonged processes of sedimentary accumulation. On the east the Colombian Basin is bounded by the submarine Beata Ridge. The ridge starts from the prominence of Hispaniola and extends to the south-west, not reaching the continental slope of South America, from which it is separated by the Aruba Passage. The ridge has the form of a narrow swell with a convex top surface and relatively gentle, almost non-dissected slopes [140]. The depths over the ridge do not exceed 2000–2200 m, and in its northern part they decrease to 1500 m.

The Venezuelan Basin is the largest in the Caribbean Sea. In its central and northern parts the depths reach 5070–5080 m. The floor here is even, whereas the remaining part of the basin is characterized by the hill-and-ridge type of relief. The ridge bounding the basin on the south is a narrow swell, with small islands, Curaçao, Bonaire and some others, rising on its top surface. To the south of the ridge, separating it from the continental slope, the small Bonaire Basin is situated with depths of up to 1800–1900 m.

On the east the Venezuelan Basin is bordered by the submarine Aves Rise lying in the year of the Lesser Antilles arc and apparently representing an older island arc that has now disappeared [141]. The top surface of the ridge lies at depths of 1000–1200 m. From the Lesser Antilles arc the Aves Rise is separated by the small Grenada Basin with a level floor and depths ranging from 2000 to 3000 m.

On the outer side of the Antillean Ridge, from Hispaniola to the median part of the Lesser Antilles arc (near Guadeloupe Island), extends the Puerto Rico Trench [135]. The floor of the trench is evened and composed of a chain of elongated narrow depressions separated by rises. The width of the floor does not exceed 2–4 miles. The slopes of the trench are steep and are diversified by steps and short lateral crests located at different levels. The average gradient of the slopes is 5–6°, but on scarps it increases up to 20° and more. The trench has it maximum depth in its western part, which, according to the data of R/V *Akademik Kurchatov*, is equal to 8395 m [65].

To the south of Guadeloupe the trench wedges out and is replaced by the broad swell-like Barbados Ridge, with Barbados Island rising on it. This ridge joins the continental slope in the region of Trinidad and is, probably, structurally connected with it. The depths over the ridge range from 400 to 1200 m. The ridge is separated from the Lesser Antilles arc by the small Tobago Trough.

The South-Antillean Zone

The Scotia Ridge forms a vast loop connecting the continents of South America and Antarctica. The northern and southern offshoots of the ridge have a block structure and are divided by 2500–3500 m deep saddle-like depressions into a series of blocks. On the northern offshoot is distinguished the Burdwood Bank with depths less than 150 m and the eastern block with South Georgia Island rising on it. On the elevated blocks of the southern offshoot rise South Orkney and South Shetland Islands [10].

The eastern section of the Scotia Ridge represents a bow-shaped swell, with volcanic South Sandwich Islands and seamounts on its upper surface (depths 1800–2000 m). Along the outer side of the southern portion of the ridge extends a second swell separated from the first one by a longitudinal depression.

The South Sandwich Trench skirts the South Sandwich Islands arc on the outer side. The slopes of the trench are steep (10–20°) and diversified by numerous steps and lateral crests (Figure 7). The floor of the trench is narrow, its width is not more than 2–4 miles. The deepest parts are represented by flat plains. The maximum depth of the trench measured by R/V *Vema* is equal to 8428 m [154]. On the south the South Sandwich Trench, as well as the ridge of the island arc, are cut off by a latitudinal fault expressed in the form of a series of scarps and narrow linear depressions.

The floor of the Scotia sea can be divided into two parts in accordance with the structure of submarine relief [10, 17]. In the eastern part submeridional strikes predominate, and in the western part sublatitudinal strikes predominate. To the east of the South Georgia Island meridian a number of bow-like swells from 500 to 1300 m in height are traced, starting at the southern offshoot of the Scotia Ridge and extending to the north. These swells seem to be the remainders of the ancient, now extinct, island arcs of the Scotia Sea. Narrow plains with depths equal to 3300–3800 m are located between them.

To the north of the South Orkney Islands a large massif rises on the sea floor, divided into a number of blocks by scarps. It is separated from the island socle by a trough with depths of up to 5400 m. Similar troughs are

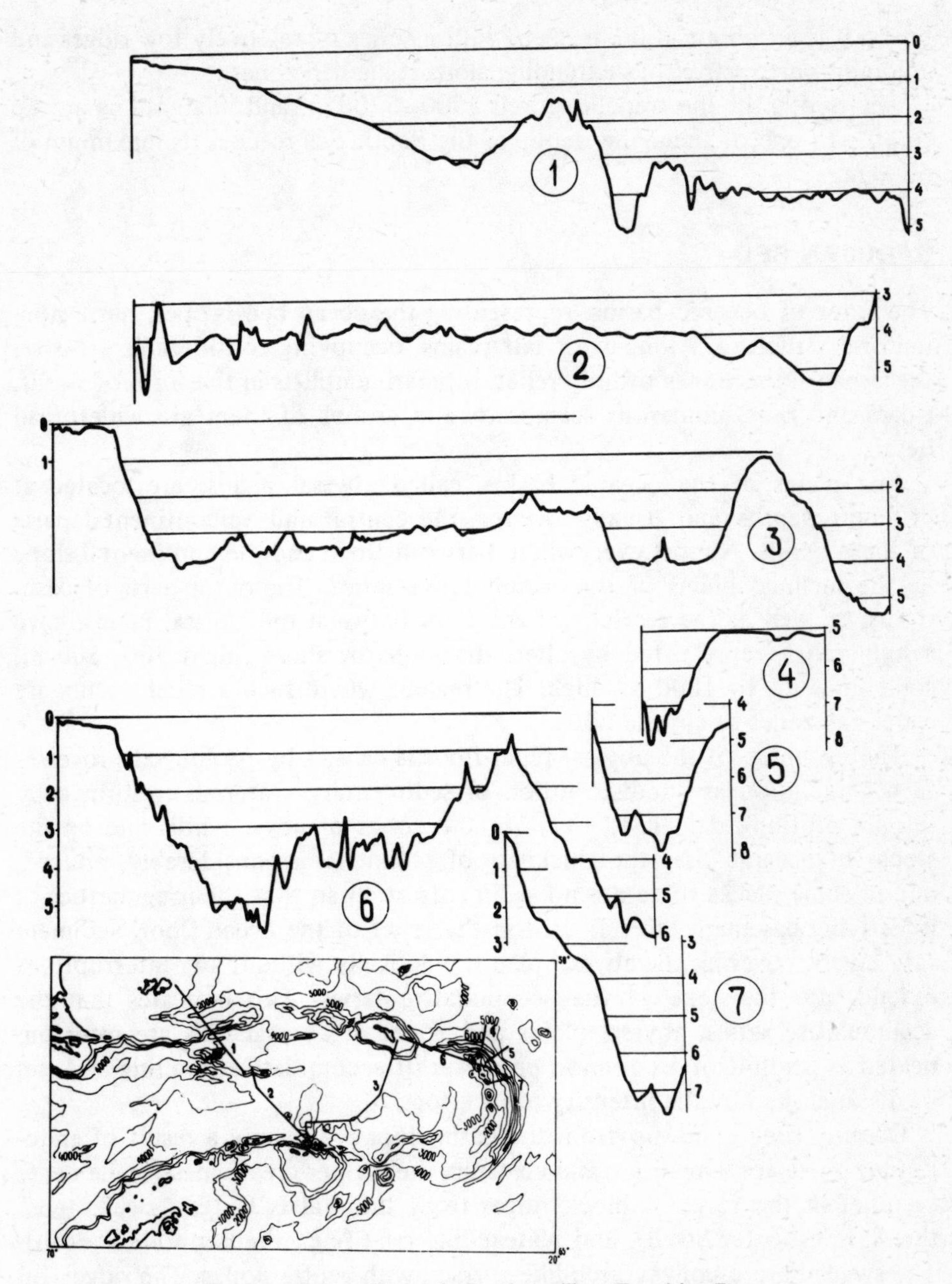

Fig. 7. Relief profiles of the South Antillean Transition Zone.

also traced in other sections of the ocean along the feet of the northern and southern offshoots of the Scotia Ridge, clearly showing the presence of longitudinal marginal faults.

The western part of the Scotia Sea up to the Drake Passage represents a hillocky plain about 4000 m deep, with a series of relatively low ridges and adjoining narrow trenches extending along it median zone.

The depths in the trenches reach 5500–5700 m, and inclinations are up to 30°. In one of them the depth of the Scotia Sea reaches its maximum of 5840 m.

5. OCEAN BED

The floor of oceanic basins, representing the ocean bed proper, has a non-uniform structure. Along with falt plains, occupying considerable areas on the ocean floor, zones of hilly relief, submarine uplifts in the form of swells, ridges and rises, numerous seamounts and groups of them are widespread here.

Flat plains of the oceanic basins, called abyssal plains, are located at different depths and usually occupy the central and epicontinental parts of these basins. Almost everywhere between them and the continental slope lie the inclined plains of the accumulative series. The outer parts of basin floors, as well as the sections of the floor between the abyssal plains, have a hilly relief, represented by alternating hills or short ridges 300–500 m, sometimes up to 1000 m, high. The regions where such a relief occurs are called the zones of abyssal hills.

The evenness of the abyssal plain floor is caused by prolonged processes of accumulation and redistribution of sedimentary material, confirmed by seismic profiling data [31, 137]. In the zones of abyssal hills and on the slopes of oceanic rises the thickness of sediments is considerably reduced, and in some places there are no sediments at all so that the irregularities of underlying basement are reflected in the relief of the ocean floor. Sedimentary layers, covering the abyssal plains, practically without any interruptions extend into the zone of the accumulative series. This indicates that the accumulative series, abyssal plains and zones of abyssal hills are interconnected as a result of a common processes of accumulative levelling and can be distinguished by the intensity of its effect.

Oceanic rises going up from the basin floor are mainly a result of endogenous processes – tectonic and volcanic, the former create massive elements of relief in the form of block ridges (e.g., the Walvis Ridge), swells (e.g., the Antilles Outer Swell), and plateau-like rises (e.g., the Bermuda Plateau). The swells have a convex arch-like surface with gentle slopes. The ridges, on the contrary, have steep, step-like slopes, their top surfaces being diversified by scarps and block uplifts. The rises have relatively level top surfaces and are in places diversified by seamounts or islands.

Another type of large uplift on the ocean floor is formed by the so-called accumulative (sedimentary) ridges, found and investigated comparatively recently. These have been created by enormous accumulations of sedimentary material as a result of prolonged action of the hydrodynamic factor [48].

The Basins of the Norwegian-Greenland Region

There are three basins here: The Greenland, the Norwegian and the Lofoten Basin that should be regarded as oceanic basins although their depths are considerably smaller than those over the Atlantic Ocean bed [54]. The Greenland Basin is located between the Mid-Oceanic Ridge and the continental slope of East Greenland. It is divided by an offshoot of this ridge into two parts: the northern (depths equal to about 3200 m) and the southern (depths equal to 3600–3800 m). In its southern part seamounts occur; the largest of them, the Vesteris seamount, reaches a height of more than 3000 m.

The Norwegian Basin, with depths of 3600–3700 m, occupies the central part of the Norwegian Sea. Against the background of an evened floor one encounters here a large number of seamounts with a height of 1000–2000 m and more, forming a zone elongated from the south to the north. According to seismic profiling data these seamounts are the tops of the ancient Aegir Mid-Oceanic Ridge half-buried under the sediments [170].

In the north-eastern part of the Norwegian Sea is located the Lofoten Basin. Its floor is represented by a flat plain slightly inclined to the west and having depths ranging from 2900 to 3200 m.

The Irminger Basin

This extends to the south-west of Iceland. Its floor is formed by an inclined plain, diversified on its edges by small seamounts and hills. The depth gradually increases to the south-west from 2400 to 3000 m.

The Labrador Basin

The Labrador Basin occupies a vast space between Greenland and Labrador. The floor of the basin is represented by a plain, slightly inclined to the south-east, with depths ranging from 3000 to 4500 m. The foot of the Greenland continental slope directly adjoins the floor of the basin, whereas along the foot of the slope of the Baffin-Greenland Rise, and of Labrador as well, there extends a well-developed plain of the accumulative series.

In the south-eastern part of the Labrador Basin is located a group of seamounts adjoining the flank of the Mid-Oceanic Ridge. In the north-western direction the heights of the seamounts gradually decrease and to the north of 58° North they entirely disappear. According to the data of geophysical investigations these seamounts are the tops of the Mid-Labrador Ridge, buried under sediments, which in the south-east is conjugated with the Mid-Atlantic Ridge [73, 125].

A characteristic feature of the Labrador Basin is the so-called North-West Mid-Ocean Canyon, morphologically expressed as a not very deep channel (going down to 100–150 m). It is traced southward along the axis of the basin and then penetrates into the Newfoundland Basin. It seems that the origin of such channels should be associated with the action of near-bottom suspension currents. This is confirmed by the fact that suspension current channels at the extensions of continental slope canyons fall into them as confluents.

At the southern extremity of the continental slope of Greenland lies the accumulative Eiric Ridge extending to the south-west and having the form of a two-pitch swell. It rises to a height of about 800–1000 m over the basin floor and gradually wedges out upwards from the foot of the slope [169].

The Newfoundland Basin

This basin occupies an intermediate position between the Labrador and the North-Western Atlantic Basins. The surface of its floor is inclined to the south, the depths vary from 4500 to 5000 m. The western part of the basin is occupied by an abyssal plain and the eastern one by a zone of abyssal hills where several large seamounts, such as the Gauss, the Miln and others more than 3000 m in height, are located [227]. From the southernmost prominence of the continental slope of the Grand Newfoundland Banks in the south-eastern direction extends the accumulative Newfoundland Ridge. It has a two-pitch slightly hilly surface, and its height gradually decreases from 1500 to 500 m.

The North-Western Atlantic Basin

The North-Western Atlantic Basin is the largest in the Atlantic Ocean. Along its northern and western edges extends a wide belt of the accumulative series whose outer boundary reaches depths of about 4500 m. Near the northern part of the Blake Plateau, in the south-eastern direction from the continental slope, extends the accumulative Blake-Bahama Ridge. Like the other ridges of this kind it also has a two-pitch slightly hilly surface, with depths over it gradually increasing to the south-east from 3500 to 5000 m. According to the data of seismic profiling, the ridge is made up of sedimentary material lying on a slight uplift of the oceanic basement [190].

Approximately in the centre of the basin is located the Bermuda Rise (Plateau) bounded by scarps or series of scarps on three sides, with the exception of its north-western side (Figure 8). This is especially pronounced on the south-eastern slope, where the height of scarps reaches 500–1000 m [168]. The surface of the plateau is flat in the west, and in the east has a hill-and-ridge relief with individual seamounts and seamount groups rising on

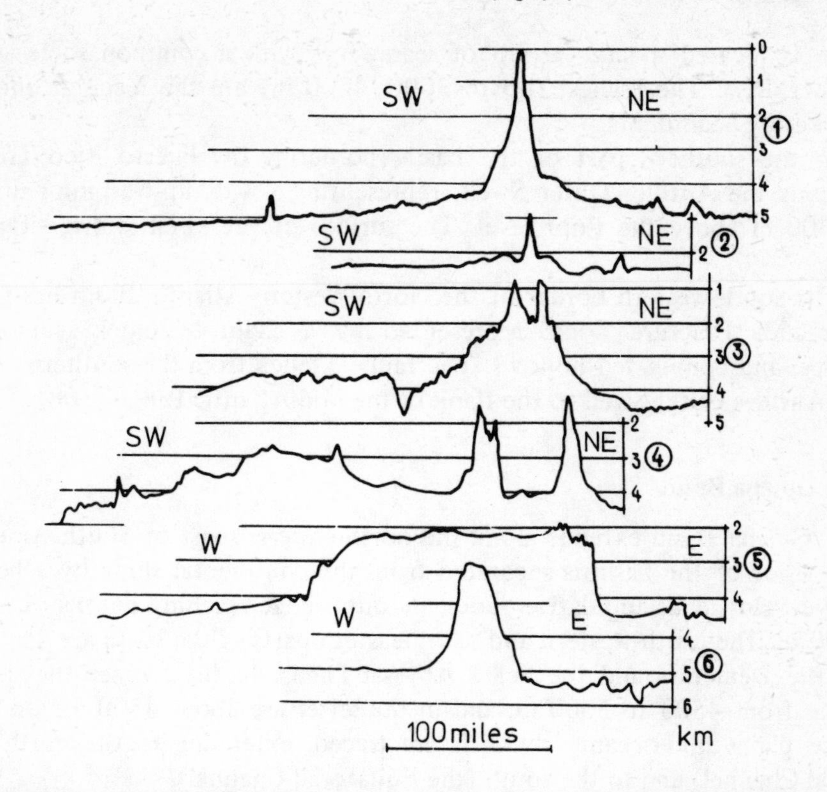

Fig. 8. Relief profiles of the oceanic rises: (1) Bermuda Plateau; (2) Seara Rise; (3) Rio Grande Plateau; (4) Sierra Leone Rise; (5) Walvis Ridge (northern part); (6) Walvis Ridge (southern part).

it. The largest is the massive Bermuda Pedestal in the centre of the plateau, on whose cut-off summit the limestone Bermuda Islands are located.

Along the north-eastern edge of the plateau the New England Seamount Chain extends to a distance of approximately 600 miles. The heights of the mountains reach 2000–3000 m. The largest of them, the Kelvin, the San Pablo, the Rehoboth, the Michael, and the Nashville Seamounts, have minimum depths from 900 to 1600 m.

From the north-east, west and south the Bermuda Rise is surrounded by the Sohm, the Hatteras and the Nares Abyssal Plains [151]. The depths in the Sohm and the Hatteras Plains are 5300–5500 m and in the Nares Plain 5800–6000 m. The monotonous flat relief of the plains is broken by separate hills and mid-oceanic canals, extending to considerable distances and connecting the plains.

Along the eastern periphery of the North-Western Atlantic Basin extends a wide zone of abyssal hills forming ridges, mainly of north-east orientation following the strike of the Mid-Atlantic Ridge. The bottoms of the valleys between the hills are usually 50–150 m below the level of the abyssal plain and the heights of the hills are 300–500 m. In the northern part of this zone is located a large group of seamounts with a common socle – the Corner Rise. The highest (up to 3000–4000 m) are the Yacutat and the Rockeway Seamounts.

In the southern part of the basin, bordering the Puerto Rico Trench, extends the Antilles Outer Swell, representing a wide up-warping rising up to 800 m above the floor level. The surface of the swell is diversified by hills.

The south-western border of the North-Western Atlantic Basin lies in the Barracuda Fracture Zone, represented by a chain of block seamounts, scarps, and conjugated hollows. The fault extends from the southern end of the Antilles Outer Swell to the flank of the Mid-Atlantic Ridge [108].

The Guiana Basin

The Guiana Basin extends along the north-eastern coast of South America. The floor of the basin is separated from the continental slope by a belt of well-developed accumulative series, its outer edge reaching depths of about 4500 m. The north-western and south-eastern parts of the basin are occupied by the Demerara and the Seara Abyssal Plains. In the former the depths range from 4500 to 5000 m, and in the latter are about 4500 m. On both these plains mid-oceanic channels are traced, extending to the north (the Vidal Channel) and to the south (the Equatorial Channel).

Between the abyssal plains is located the Seara Rise representing a plateau rising 200–500 m above the basin floor and bounded by scarps. The surface of the plateau is diversified by a group of seamounts up to 2500–3000 m high.

In the south-eastern part of the Guiana Basin there is also a large number of seamounts, with heights up to 3000–3500 m. They are mainly arranged in the form of chains, extending from the continental slope to the east along the equator at latitudes 1°35′ and 4° South. On tops of the two largest seamounts rise small volcanic islands – the Rocas and the Fernando de Noronha. A range of seamounts with merged basements forms the so-called North-Brazilian Ridge [150].

The Brazilian Basin

Along the continental slope of Brazil extends the belt of a well-developed accumulative series. Its outer edge reaches depths of 4500–4800 m where it passes smoothly into the surface of abyssal plains. The largest of these plains – the Pernambuco Plain – is located in the northern part of the basin. Its surface is slightly inclined to the south, the depths ranging from 5200 to 5600 m. Further to the south abyssal plains of smaller size are found, separated by zones of rolling floor. The eastern part of the Brazilian Basin is occupied by a wide zone of abyssal hills 300–500 m high, elongated in the form of ridges to the north.

There is a large number of seamounts on the floor of the Brazilian Basin, some of them scattered at random but the majority being concentrated along the latitudinal zones of 8°30′ and 13° South and on the extensions of the northern and southern slopes of the Abrolhos Bank. Near this bank are situated the largest seamounts, such as the Rodgers, the Morgan, the Hotspur, the Vittoria, the Jazzer, and the Davis with minimum depths over them equal to 35–45 m and their tops cut off as a result of sea abrasion. Almost in the centre of the basin there is a large rise, crowned with small volcanic islands – Trinidade and Martin Vaz.

The Rio Grande Rise

This massive structure, separating the Brazilian and the Argentine Basins, consists of two parts joined by a common socle with depths of about 4000 m. The western part of the rise is formed by a step-and-block plateau, bounded by scarps on every side. The plateau surface, having depths from 2500 to 2900 m, is diversified by seamounts up to 1500–2000 m high (see Figure 8). To the east of this plateau stretches a meridionally elongated block ridge with depths over it equal to 2500–3000 m [186].

The Argentine Basin

The structure of its floor is rather simple, the depths gradually increase southwards from 4800 to 6000 m. Here a band of accumulative series extends along the continental slope, wide in the north and narrow in the south, the depths of its outer edge increasing southwards from 4500 to 5500 m. Along it lies the relatively narrow Argentine Abyssal Plain with a flat floor, slightly inclined to the south. The central part of the Argentine Basin has a gently rolling relief with occasional hills. The eastern part of the basin represents a zone of continuous abyssal hills 500–800 m high. One notes the ridge-like arrangement of hills generally orientated along the strike of the Mid-Atlantic Ridge. The southern boundary of the basin is formed by the Falkland Ridge, extending eastwards on the continuation of the northern scarp of the marginal Falkland Plateau [186].

On the outer side of the South Sandwich Trench extends the South Antilles Outer Ridge elevated above the basin floor to a height of approximately 500 m. Its surface is diversified by ridges of hills, as well as scarps and small block seamounts.

The African-Antarctican Basin

This extends over a considerable distance along the submarine boundary of Antarctica. On the west it includes the deep-water part of the Weddell Sea, where depths are equal to 4800–4900 m. In the eastward direction the depth of the basin gradually increases, reaching 5300–5400 m. Along the continental slope of Antarctica extends a belt of accumulative series, most developed in the Weddell Sea. To the north of Queen Maud Land in the zone of this series lies the rounded Maud Rise representing a block plateau elevated approximately 500 m above the ocean floor [26].

The Iceland Basin

This is located to the south of Iceland, being as it were the analogue of the Irminger Basin on the other side of the Reykjanes Ridge. Its depths gradually increase to the south-west from 2200 to 3000 m. The floor of the basin is evened. Along its south-eastern border lies the mid-oceanic Maury Channel penetrating on the south through the zone of abyssal hills into the North-Eastern Atlantic Basin [169].

The North-Eastern Atlantic Basin

For the most part the floor of the basin is level and forms the Biscay Abyssal Plain. Its depths are 4500–4800 m, and the floor is slightly inclined to the south. On the western side the abyssal plain is skirted by a relatively narrow zone of abyssal hills generally scattered at random [180].

In the north-western part of the basin there are groups of seamounts up to 1500 m high. In the southern part of the basin there are also seamounts, such as the Biscay, the Cantabria and others, up to 2000 m high. Their block structure and the findings of sedimentary and metamorphic rocks on them indicate that the seamounts are *fragments* of the broken-up continental margin.

The Iberian Basin

The Iberian Basin is located to the west of the Iberian Peninsula. Almost all the basin is occupied by an abyssal plain 5100–5200 m deep with isolated hills and seamounts on it. Along the western and southern borders of the plain extends a narrow zone of abyssal hills, on the east the basin floor directly adjoins the foot of the continental slope.

The Azores-Gibraltar Rise Zone

This includes the Azores-Cape St Vincent Ridge and groups of seamounts to the west of the Strait of Gibraltar. This zone is tectonically very significant, serving as a link between the Mid-Atlantic Ridge and the Mediterranean (Alpine) Belt.

The largest seamounts are located in the eastern part of the zone. They form a horseshoe-shaped chain, convex to the west and known as the Horseshoe Rise, which includes the well-known Gorringe, Josefine, Ampere, and other seamounts with their heights reaching more than 4500 m and minimum depths ranging from 40 to 150 m [151]. To the south-east of the Horseshoe Rise are the volcanic Madeira Islands, the Seine Seamount and other rises. To the north of the Horseshoe seamounts group a chain of mountains is located, which together form the Iberian Ridge elongated in the north-eastern direction [20].

Between the Horseshoe Rise and the Azores volcanic massif extends a rise, elevated above the basin floor to approximately 400–500 m. Its surface is diversified by numerous hills and small seamounts, and the northern and southern slopes are represented by scarps formed along the lines of latitudinal fractures [179].

The Canary Basin

The Canary Basin is a large basin of the Atlantic Ocean. The accumulative series here is very well developed and reaches a width of up to 250 miles. Within it one finds individual seamounts, and the volcanic Canary Islands are located here. These islands form a chain passing along an arc which is convex to the south. The heights of the islands are more than 2000 m, the feet of submarine slopes reach depths of 3200–3500 m. In the saddles between the islands, however, the depths do not exceed 2200 m, which indicates the presence of a common socle.

The eastern part of the Carnary Basin is occupied by an abyssal plain whose floor is slightly inclined to the south, and the depths gradually increase from 5000 to 6000 m. The western part of the basin represents a wide zone of continuous abyssal hills. They are 500–800 m high and are mainly arranged in the form of ridges with submeridional strike [209].

The Cape Verde Basin

On the north-east the basin is bounded by a belt of the accumulative series, which near the Cape Verde Islands merges with the islands series and reaches a width of almost 500 miles. Suspension current channels are in places encountered on the extension of continental slope canyons. The Cape Verde Islands form a horseshoe-shaped chain, convex towards the east. The height of the islands amounts to 1000–2000 m and more. In the straits between the islands the depths vary from 200 to 2800 m, and the feet of submarine slopes reach depths of 3000–3500 m.

To the south-west of the islands is located the small Gambia Abyssal Plain with depths equal to 5000–5500 m. The remaining part of the basin floor is occupied by a wide zone of abyssal hills. Within it there are individual seamounts, the largest of which, the Krylov Seamount, reaches a height of 3200 m.

The Sierra Leone Rise

The Sierra Leone Rise represents a large elevation of the ocean floor, slightly elongated from the south to the north and formed by a *stepped* plateau with depths of 3500–4000 m [95]. In the northern part of the plateau a group of seamounts rises up to 1500–2000 m (see Figure 8). The steep scarp of the northern slope of the plateau is located on the extension of the Guinea Fracture and is, probably, genetically connected with it.

The Sierra Leone Basin

The Sierra Leone Basin lies to the south-west of the rise of the same name. Most of its floor is occupied by an abyssal plain, where the depths are equal to 4800–5200 m, but in the southernmost part increase to almost 6000 m. The accumulative series belt in the north and the zone of abyssal hills in the south are poorly developed.

The Guinea Basin

This occupies the western part of the Gulf of Guinea, while in its eastern part is situated a wide accumulative series zone bounding the alluvial cone of the River Niger. From the Cameroun volcano to the south-west one can trace a chain of small volcanic islands, located within the accumulative series zone. These are the Fernando Poo, the Principé, the São Tomé, and the Annobon Islands. The feet of the submarine slopes of these islands reach depths of 2700–3000 m, and the height of the largest island is more than 2000 m above the ocean level.

Most of the Guinea Basin floor is occupied by an abyssal plain with depths equal to 5000–5100 m, where individual hills and small seamounts are encountered. Along the southern boundary of the basin a narrow belt of abyssal hills is traced [60].

In the south-east the basin is bounded by the wide, gently sloping Guinea Rise with depths over it equal to 4700–4800 m. The surface of the rise is undulating and diversified by individual seamounts up to 2000–2500 m in height.

The Angola Basin

Along its eastern edge extends a well-developed zone of accumulative series, its outer edge reaching depths of 4600–4800 m. A great part of the basin floor is occupied by an abyssal plain whose surface is slightly inclined to the west, with depths gradually increasing from 4800 to 5600 m. Along the western edge of the basin extends a relatively narrow zone of abyssal hills, their heights reaching 300–500 m and the bottoms of valleys between the hills lying 100–200 m below the basin floor level.

On the floor of the Angola Basin there are numerous individual seamounts and groups of them. The most significant group of these seamounts is located in the northern part of the basin, and most of them are concentrated along the line connecting the Cameroun volcano with the Saint Helena Island. The largest of them are the Shirshov, the VNIRO and other seamounts, with heights reaching more than 5000 m and minimum depths over them of about 400 m [3, 60].

The Walvis Ridge

The Walvis Ridge extends in the south-western direction from the continental slope of Africa to a distance of approximately 1000 miles. All along its length the ridge represents a block structure with a plateau-like top surface and step-like slopes (see Figure 8). On the south-east the slope inclination reaches 12–15°, and on the north-western slope it is not greater than 6°, on individual scarps up to 10°. This imparts a distinct asymmetry to the ridge.

The apical plateau of the ridge is divided into three large blocks by two saddle-like depressions. The north-eastern block adjoining the continental slope has depths of 2300–2500 m. The median block is mainly located at depths of 2000–2500 m, but in the central part there is a rise on it with a minimum depth equal to 216 m, known as the Valdivia Bank. The south-western block represents a rather narrow massive swell, with depths over it ranging from 900 to 2000 m. From its southern end in the direction of the Tristan da Cunha Islands extends a zone of well-developed uplifted fault-blocks and volcanic seamounts, the largest of them being the Wüst Seamount [91].

The Cape Basin

The Cape Basin is situated to the south of the Walvis Ridge. Along its eastern border extends a rather well developed zone of accumulative series, whose outer edge reaches depths of about 4500 m. The Cape Abyssal Plain is a relatively narrow strip of flat floor with depths ranging from 4600 to 5200 m. The rest of the Cape Basin Floor is occupied by a wide zone of abyssal hills, where the general level of depths reaches 5300–5400 m [42].

On the floor of the Cape Basin there are numerous seamounts. The largest of them, the Vema, the Discovery, the Meteor and others, are as high as 4000 m and more. Most of the large seamounts are concentrated along lines with a north-east strike.

The Agulhas Basin

The floor of the basin is 5200–5400 m deep. A considerable part of it is occupied by a zone of abyssal hills. A narrow strip of accumulative series and abyssal plain extends only along the foot of the continental slope. In the north-eastern part of the basin lies the rather large plateau-like Agulhas Rise with depths over it equal to 2000–2500 m.

6. MID-OCEANIC RIDGE

The Mid-Oceanic Ridge is clearly defined throughout the Atlantic Ocean and the Norwegian-Greenland Basin and has characteristic structural features. Its relief is complicated, formed by valleys, scarps, transverse trenches, individual seamounts, and disruptured plateaus stretching along its strike. All this diversity of the forms of relief lies within three basic geomorphological zones: the crest (rift) zone and the flanks of the ridge on both sides of it.

The crest zone is 1000–1500 m above the ridge surface and considerably more dissected. Almost everywhere a series of deep and narrow rift valleys (the 'median valley') extends along its axis. These are usually regarded as graben-like structures that have been formed as a result of the spreading of the Earth's crust. Geomorphological studies, including those performed by R/V *Akademik Kurchatov* [61], are indicative of an intermittent *en echelon* arrangement of the rift valleys. On both sides they are bounded by series of rift ridges (the 'valley walls'), dissected plateaus, and individual seamounts (Figure 9). The tops of crests and seamounts are mainly peaked, the slopes are steep (10–20°), the bottoms of inter-ridge valleys are narrow and concave, flat sections are practically absent. All this indicates the recent formation of the rift zone relief and the block-and-fault character of its structure.

The flanks of the Mid-Atlantic Ridge constitute gradually descending

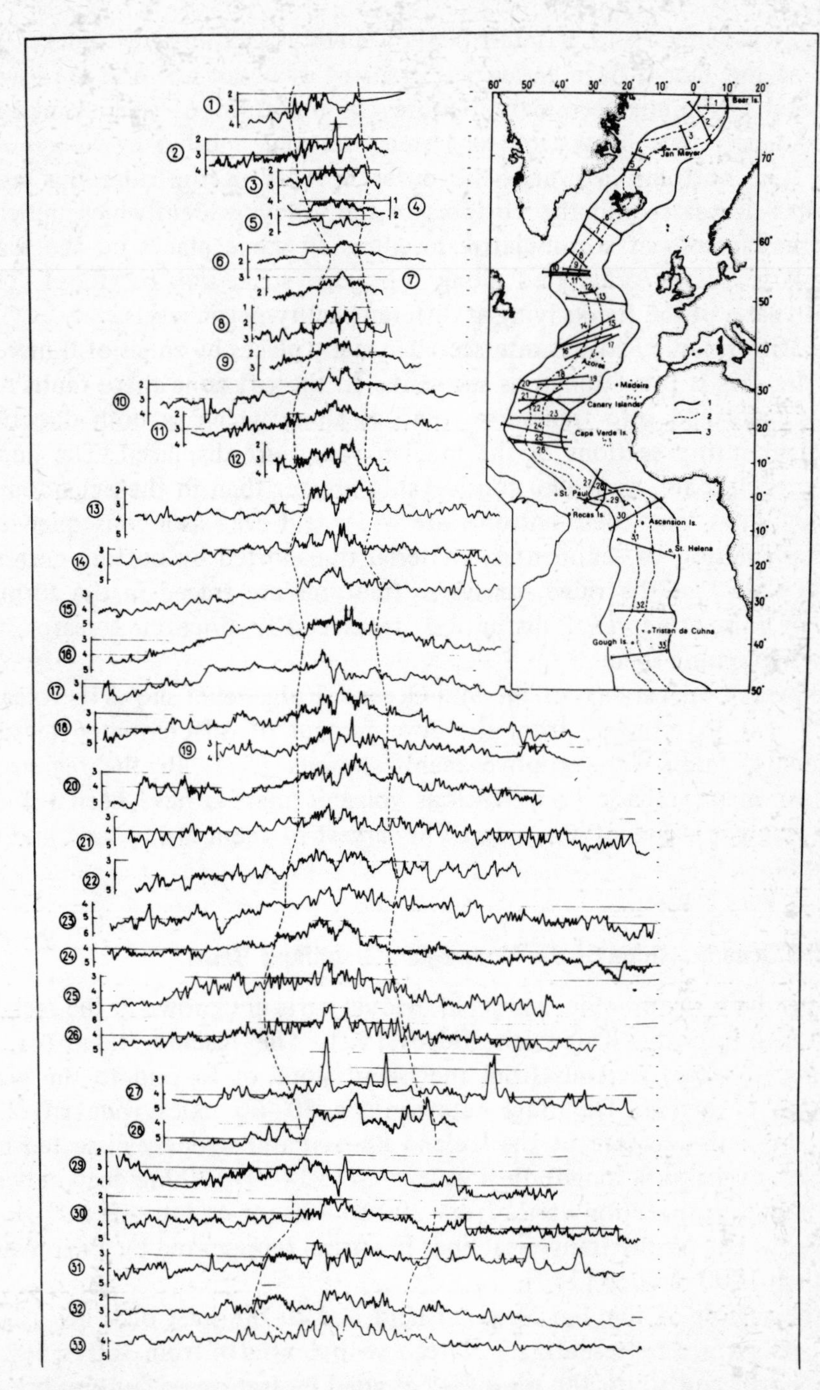

Fig. 9. Relief profiles of the Mid-Oceanic Ridge in the Atlantic Ocean and of the Norwegian-Greenland Basin. The rift zone is marked by dash line. (1) ridge boundary; (2) rift zone boundary; (3) location of profiles.

surfaces, dissected into a multitude of mainly assymmetric blocks. The slopes of the blocks have an average gradient of about 10° and are usually diversified by minor secondary features in the form of small banks and pointed peaks. The blocks are separated from one another by valleys with mainly flat bottoms. Towards the outer borders of the ridge the height of blocks decreases and the bottoms of the valleys widen, which indicates an increasing role of accumulative levelling. In some places on the flanks rather large scarps orientated along the ridge strike can be traced. They separate parts of the flanks lying at different bathymetric levels.

The Mid-Oceanic Ridge is intersected in many places by zones of transverse faults dividing it into numerous segments. In the rift zone these faults have been transformed into transverse trenches along which in both directions the neighbouring sections of the median valley are displaced. The depths in the trenches are, as a rule, considerably greater than in the neighbouring median valleys, and their bottoms are wider and even as a consequence of the accumulation of sedimentary material transported by bottom currents. On the flanks of the ridge transverse fractures are traced in the form of zones of submarine relief disruption, transversally orientated scarps, and trenches adjoining them.

The overall appearance of the Mid-Oceanic Ridge relief shows its volcano-tectonic nature resulting from the development of differentiated tectonic movements, faults and eruptive manifestations [57]. In the regions of the most massive basic lava effusions volcanic massifs have been formed, diversfying the relief of the ridge. The largest of them are Iceland and the Azores.

The Mid-Oceanic Ridge of the Norwegian-Greenland Basin

The ridge here is not wide, and its individual parts are known as the Iceland, the Mohns, and the Knipovich Ridges [52]. The Iceland Ridge (or the Kolbeinsey Ridge) extends from the island slope of Iceland to the north reaching 71° North. The ridge represents a 50–60 miles wide rift zone raised above the surface of the Iceland Plateau and strongly dissected into numerous crests with longitudinal valleys up to 1400–1700 m deep between them. Vertical dissection varies from 200 to 700 m. At latitude 69° North the ridge is cut by the transversal Spar Fracture, represented by a depression more than 1700 m deep [170].

In the region of the Jan Mayen Island and to the west of it lies a large massif, its surface representing a plateau with depths of from 300 to 600 m. To the north and south the massif is bounded by transverse faults expressed by steep scarps and conjugated trenches. The northern trench is larger, its depth varies from 2000 to 3800 m. It separates the Jan Mayen Island from the Mohns Ridge [452]. The total displacement of rift structures along the two fractures amounts to 120 miles.

The Mohns Ridge extends in the north-eastern direction approximately to 10° East, where it turns to the north and then passes into the Knipovich Ridge. The depth of dissection on both ridges is equal to 500–1000 m, occasionally reaching 1500 m. The median valley on the Mohns Ridge is represented by several narrow valleys where the depths reach 2800–3400 m, and on the Knipovich Ridge the valley extends in the form of a single trench more than 300 miles long and up to 3300–3500 m deep. The highest peaks of the Mohns and the Knipovich Ridges lie 600–1000 m below the sea level. The inclination of slopes on the average amounts to 10°, in individual parts reaching 25–30°, particularly on the slopes of rift valleys [52].

The Rift Zone of Iceland

This is a link between the rift zones of the Mid-Oceanic Ridge of the Norwegian-Greenland Basin and the Mid-Atlantic Ridge and extends through the zone of the Central Graben of Iceland, crossing the island obliquely from the north-east to the south-west. On both sides of it there are zones of the development of Tertiary plateau basalts, similar to those occurring in the Faeroes, Greenland and Scotland but not so old. [8]. The Central Graben of Iceland is filled up with Quaternary and recent volcanic formations and characterized by present-day tectonic activity. Almost all the epicentres of earthquakes recorded in Iceland and all the active volcanoes are concentrated here [222]. The relief of this zone is represented by elongated peaked ridges, chains of volcanic cones, and tectonic valleys that can be regarded as analogues of rift valleys and ridges of the Mid-Oceanic Ridge.

Iceland is surrounded by an island shelf, its width reaching 50–80 miles and only in the south decreasing to 8–10 miles. The shelf is almost everywhere dissected by radial submarine valleys into a series of banks. The depths on these banks are equal to 150–180 m, in the valleys reaching 250–300 m and at the northern coast 400–500 m [51]. On the shelf of the northern coast of Iceland, where structures of the Central Graben outcrop, there is a noticeable displacement of submarine relief forms to the west. The distribution of earthquake epicentres in this region indicates the presence of the transverse Tjornes Fracture. To the north of the fault, starting from the small Kolbeinsey Island, extends a chain of narrow elongated banks and ridges, which then merge with the Iceland Ridge. On the south-western coast of Iceland from Cape Reykjanes another chain of submarine rocks and ridges extends across the shelf, further on passing into the Reykjanes Ridge [52], the northern part of the Mid-Atlantic Ridge.

The Reykjanes Ridge

The Reykjanes Ridge extends from the shelf of Iceland to the south-west down to latitutde 53° North, where it is bounded by the latitudinal Gibbs Fracture Zone. The ridge has clearly defined rift zone and flanks, but their structure changes from north to south [41, 226]. The total width of the ridge gradually increases to the south-west from 300 to 500 miles, and the width of the rift zone increases from 30 to 80–100 miles.

The rift zone near Iceland represents a massive swell bounded on both sides by steep (up to 20–25°) scarps 600–650 m high. Its top surface is situated at depths equal to 900–1000 m and is dissected into numerous peaked crests and chains of small seamounts. There is no typical median valley here, but individual deeper linear depressions (going down to 1100–1200 m) may be regarded as a kind of rudimentary form of this valley. On each side of the rift zone lie gentle undulating plains forming the flanks of the northern part of the Reykjanes Ridge. Steep scarps up to 500 m high divide them into two levels lying at depths equal to 1500–1600 and 2000–2500 m respectively. In some places these plains are diversified by small seamounts and ridges.

To the south of latitude 60° North the relief of the Reykjanes Ridge gradually changes. The rift zone becomes wider, dissection of the relief increases. To the south of latitutde 58° North a well expressed deep median valley appears, with depths over its bottom reaching 2500–2900 m, while over the bordering crests the depths are 1600–1800 m. These crests are formed by chains of asymmetric blocks more than 500 m high. At latitudes 57° and 55° North the rift zone is cut by transverse depressions along which a relatively small displacement of rift structures is seen. The flanks of the southern part of the Reykjanes Ridge are situated at depths ranging from 2200 to 2800 m. Their surface is represented by a continuous alternation of blocks stretching along the strike of the ridge. Individual conic seamounts more than 1000 m high are also encountered.

The Gibbs Fracture Zone is represented by a double series of trenches cutting the rift zone and going partly into the flanks [203]. Most developed is the northern trench where the depths reach 4000–4200 m. Overall displacement of rift structures along the system of both faults reaches there about 200 miles.

The North-Atlantic Ridge

The ridge is divided by the Azores Volcanic Massif into two parts: the northern and the southern. The northern part of the ridge is 600–800 miles wide and in plan forms an arc convex to the east. The rift zone is well expressed here and represents a massive swell 90–110 miles wide raised above the flanks by 1000–1500 m. Vertical dissection is mainly equal to 500–800 m, but near the rift valley it increases considerably. The bottoms of rift valleys are situated at depths ranging from 2500 to 4000 m, the tops of bordering crests rising up to the depth level of 1800–2000 m. The highest seamounts have minimum depths of less than 1000 m over them. Their tops are peaked, the slopes are steep (up to 20°).

The northern part of the North-Atlantic Ridge is cut in many places by relatively small transverse fractures. The best investigated are the Faraday (at 49° N), Maxwell (at 48° N), Humboldt (at 42°15′ N), and Kurchatov (at 41° N) Fracture Zones, along each of them a displacement of rift structure by 15–20 and more miles is observed.

The flanks of the ridge are represented here by surfaces, gently descending on both sides, with depths ranging from 3000–3500 to 4500–5000 m. In some places one sees here local longitudinal scarps dividing the flanks into a series of steps. The relief of the flanks is generally formed by an alternation of blocks, stretched along the strike of the ridge and inclined towards the rift zone, which makes their profile asymmetric. Against this background individual seamounts are encountered. The largest of them are the Altair on the western flank and the Ant-Altair on the eastern flank of the ridge with heights equal to more than 2500 m.

To the north-east of the Azores Massif the flank of the North-Atlantic Ridge is diversified by large obliquely transversal sturctures, such as the Palmer and the Mesjatzev Ridges.The former extends from the north-west to the south-east approximately to a distance of 140 miles and is cut by the longitudinal King's Trough into two crests. The flat bottom of the trough reaches depths ranging from 4200 to 5300 m. The depths over the crests do not exceed 2500 m [180]. In the south-east the Palmer Ridge is bounded by the perpendicularly situated Mesjatzev Ridge [20]. The latter represents a chain of crests and seamounts extending to a distance of up to 400 miles. The foot of the ridge is located at depths equal to about 4000 m, the tops rising up to a depth level of 1800–2000 m.

The volcanic Azores Massif has a complicated structure. Its highest mountains rise over the ocean level in the form of a group of islands. The two western islands – Flores and Corvo – are situated within the rift zone and are separated from the others by a rift valley. To the south of these islands the rift zone is cut by the transverse West Azores Fracture Zone having the form of a deep trench. The principal group of the Azores, including São Miguel, Terceira, São Jorge and the other islands, rises on a massive plateau situated eastward of the rift zone. The submarine socles of the islands are connected by ridges orientated to the east-south-east. The trenches separating the ridges reach depths ranging from 1200 to 3100 m. On the north and south the Azores Plateau is bordered by stepwise scarps, stretching along the strike of the East Azores Fracture Zone. The scarps are also traced along the Azores-Cape St Vincent Ridge [179].

To the south of the Azores Massif the North-Atlantic Ridge extends in the form of an enormous arc convex towards the west. Here the rift zone is well expressed, its width gradually increasing to the south from 80 to 160 miles. The total width of the ridge from the Azores to latitude 15° North is equal to 900–1000 miles, contracting to the south to 300–400 miles [32].

All the way from the Azores to the equator the ridge is dissected by numerous transverse fractures. Large fractures are traced not only in the rift zone but also on the flanks right up to the outer boundaries of the ridge. To the north of 15° North they are expressed on the flanks mainly in the form of a zone of submarine relief disruption, to the south of it mainly along series of steep scarps bordering the segments of the ridge displaced with respect to one another. The best investigated are the Oceanographer (at 35° N), the Atlantis (at 30° N), the Kane (at 24° N), the Cape Verde (at 14°30′ N), the Vernadskii (at 7–8° N), São Paolo (at 1°30′ N), the Romanche (at the equator), and the Chain (at 1°30′ S) Fracture Zones. The displacement of rift structures observed on all of them ranges from 30–40 to 250–300 miles. Maximum displacements are associated with the Vernadskii, the São Paolo, and the Romanche Fracture Zones [19, 61, 86, 122, 139, 153, 230]. The depths in the transverse trenches vary from 4500 to 5200 m, in the Romanche Trench reaching a maximum value for the Mid-Atlantic Ridge of 7856 m. The slopes of trenches are steep (up to 20–30°) and diversified by steps. The crests, bordering the trenches in the north and south as a rule have an asymmetric profile, i.e., their inner slopes are steeper than the outer slopes. Longitudinal rift structures dissect these crests into blocks, a good example of which is the Atlantis Fracture Zone (Figure 10).

A system of rift valleys throughout the ridge to the south of the Azores is represented by clearly defined narrow and deep valleys arranged in succession one after another or en echelon. To the north of the Atlantis Fracture Zone the depths of the bottoms of rift valleys are about 3000 m, increasing to 3600–4000 m to the south of it. The rift ridges are formed by chains of elongated blocks with predominant depths over them of 1800–2500 m, but individual seamount tops rise to the depth level equal to less than 1000 m (Figure 11). Vertical dissection of the rift zone is gradually increasing to the south, ranging from 800 to 1500 m [193].

The flanks of the southern part of the North-Atlantic Ridge have a characteristic block-and-ridge dissection with orientation along the strike of the ridge, although in the transverse fault zones it can be disturbed. The widths of blocks are equal to 10–20 miles, their heights range from 300 to 800 m.

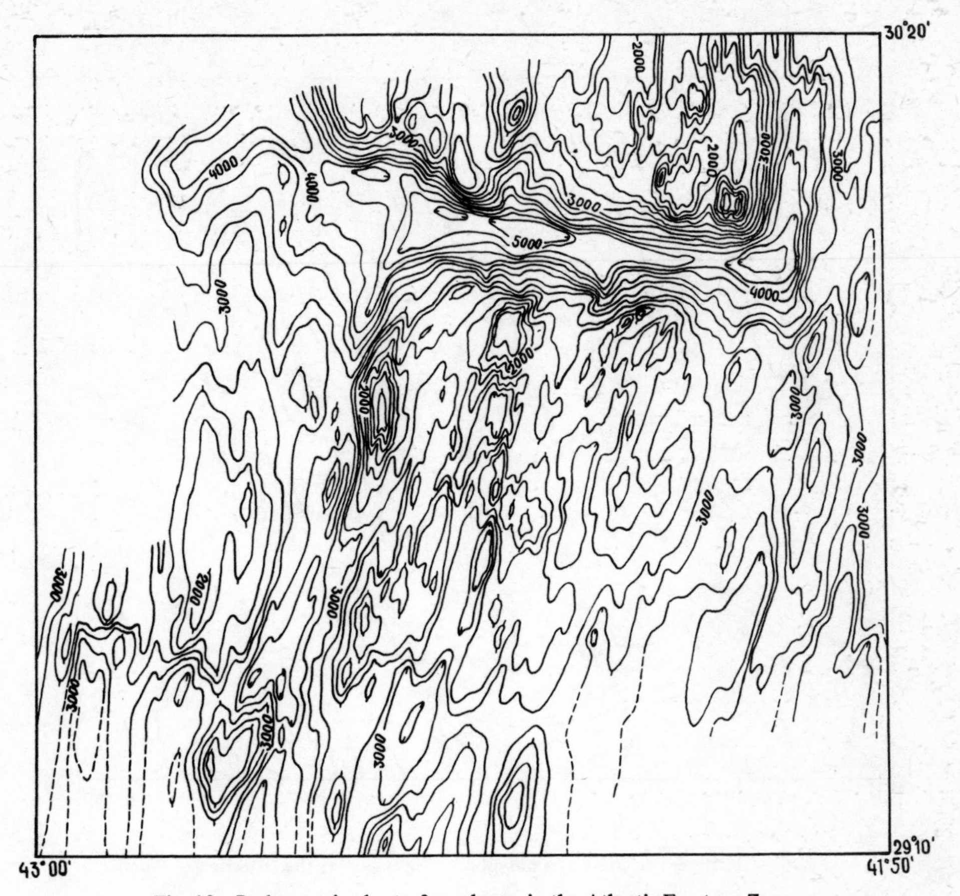

Fig. 10. Bathymetric chart of a polygon in the Atlantis Fracture Zone.

Here and there seamounts with altitudes from 1500 to 4000 m rise over the flanks. On the western flank low seamounts predominate and on the eastern flank to the south of the Azores there is a group of large seamounts. These form a chain extending to the south-east and include such seamounts as the Atlantis, the Plato, the Cruiser, the Great Meteor, and others [151, 206]. The slopes of the seamounts are steep (more than 20°) and their summits are located at depths of from 265 to 375 m topped by marine abrasion when the ocean level was much lower in the Plio-Pleistocene time.

The South-Atlantic Ridge

All over the length of the ridge from the Chain Fracture Zone to Bouvet Island a well-developed rift zone is traced. Its width gradually decreases to the south from 250 to 120 miles, and the total width of the South-Atlantic Ridge is reduced from 1000 to 500 miles. To the south of the

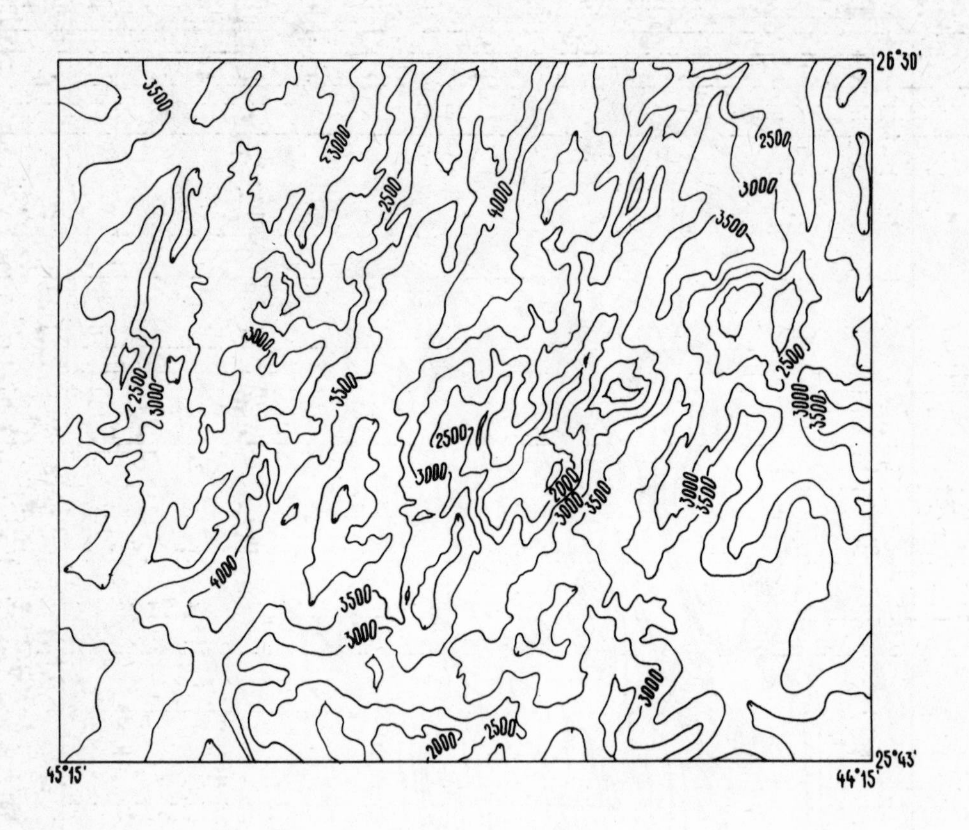

Fig. 11. Bathymetric chart of a polygon in the region at the latitude 26° North.

Chain Fracture Zone there is no clearly defined rift valley. However, near Ascension Island a deep rift valley appears, its bottom reaching a depth of more than 4000 m [60]. The tops of rift ridges are located at depths of 2500–2700 m, and individual seamounts have minimum depths of less than 1000 m. One of the large seamounts rising from a depth of 3800 m appears above the ocean level as Ascension Island. Here the ridge is cut by a large transverse fracture, along which the displacement of rift structures reaches 120 miles [231].

To the south the rift zone on the whole has an arch-like shape and is dissected by deep rift valleys and parallel valleys into series of ridges. The depths in the rift valleys reach 4000–4300 m. The tops of the crests are situated at depths of 2500–2800 m, individual seamounts rising higher. Vertical dissection in the rift zone varies within 500–1000 m. At this section of the ridge there are several large transverse fractures, the best known of which are: the Saint Helena (at 17° South), the Hotspur (at 19° South), the Martin Vaz (at 22° South), the Rio Grande (at 30° South), and the Gough (at 40° South) Fracture Zones, but displacements of rift structures along them are comparatively small [34].

The southernmost part of the rift zone of the South-Atlantic Ridge is expressed less distinctly. The rift valley here is traced in the form of separate segments, where the depth of its bottom reaches 3500–4000 m. The rift ridges are situated at depths of not more than 2500 m. Vertical dissection ranges from 400 to 600 m. At latitude 46° South is traced a transverse fracture on the continuation of the Falkland Fracture Zone. Near Bouvet Island dissection of the rift zone increases sharply. Here the ridge is cut by the north-east orientated large Bouvet Fracture in the form of a trench with depths equal to 5000–5200 m, and the displacement of rift structures reaches 140 miles. The large Captain Speiss Seamount rises to the west of the trench, and to the east of it – a whole massif consisting of several seamounts, one of their summits forming the volcanic Bouvet Island [171].

The flanks of the South-Atlantic Ridge all along its length have a rather uniform structure. Their surface is formed by series of ridges consisting of asymmetric blocks 300–500 m high. The western flank is on the whole less dissected than the eastern one and seamounts on the former are rare. On the eastern flank many large seamounts are known. In its northern part lies the volcanic Kurchatov Seamount more than 3000 m high, in the central part is the volcanic massif of St Helena Island more than 5000 m high, and further to the South is a group of mountains forming the volcanic Tristan da Cunha Islands, then Gough Island, and the Crowford, Macnish, and Admiral Zenker Seamounts more than 3500 m high.

The Atlantic-Antarctic Ridge

This is the intermediate link between the Mid-Oceanic Ridges of the Atlantic and the Indian Oceans. The width of the rift zone here is about 100 miles and the total width of the ridge is not more than 500 miles. The ridge is cut by a series of transverse faults orientated to the north-east and parallel to the Bouvet Fracture. Along the ridge axis extends a series of rift valleys, the depths in them exceeding 4000 m. The tops of bordering rift ridges rise to the depth level of less than 2500 m. The flanks of the ridge are formed by surfaces progressively descending on both sides and dissected by ridges stretching along the strike.

7. SYMMETRY AND ASYMMETRY OF THE OCEAN FLOOR RELIEF

When one examines the general scheme of the structure of the Atlantic Ocean floor relief, the first thing that attracts attention is its symmetry relative to the axis of the Mid-Oceanic Ridge, which has been previously noted by other authors as well [34, 151]. Morphologically this axis is expressed by a series of rift valleys, bounded by rift ridges. As a rule, the height and the size of blocks comprising these ridges are approximately the same on both sides of the rift valley and show a regular decrease as the distance from the ridge axis increases. Secondary dissection, diversifying the ridge slopes, however, masks considerably the symmetry in the structure of large and medium-size forms of relief. Besides this, additional disturbances are created by transverse faults. Only on profiles strictly perpendicular to the ridge axis and not crossing the zones of transverse fractures does this symmetry manifest itself quite distinctly. The symmetry is expressed not only in the regular alternation of ridges and a progressive decrease in their dimensions, but also in the shape of the blocks comprising them. Their slopes facing the rift valleys are steeper than the opposite slopes. In the western part of the ridge the blocks thus seem to be inclined to the east and in the eastern part they are inclined to the west.

On both sides of the Mid-Oceanic Ridge oceanic basins are located in whose floor relief structure one can also note a certain symmetry relative to the rift zone axis. Firstly, the very location of the basins is to a certain extent indicative of it. The Irminger Basin lies opposite the Iceland Basin, the Labrador Basin – opposite the North-Eastern Atlantic Basin, the North-Western Atlantic Basin – opposite the Canary Basin, the Brazillian Basin – opposite the Angola Basin, the Argentine Basin – opposite the Cape Basin. Approximately opposite one another on both sides of the ridge are many of the transverse rises separating the basins. But most of all this symmetry manifests itself in the character of submarine relief. Everywhere those parts of the floor of the basins that adjoin the Mid-Oceanic Ridge are occupied by the zones of abyssal hills which are then replaced by flat abyssal plains. In the structure of the relief of abyssal hills there is a certain resemblance to the relief of the flanks of the Mid-Oceanic Ridge. A large part of the hills is arranged in the form of ridges orientated along the strike of the ridge, and the hills themselves represent blocks with asymmetric profile.

Along the coasts of the continents surrounding the Atlantic Ocean there is a strict succession of zones: the shelf, the continental slope and the continental rise bounding the ocean bed on the west and the east. In this, one can also notice a certain degree of symmetry in the structure of submarine relief, which is emphasized by the similarity in the outlines of continental margins of North and South America, on the one hand, and those of Europe and Africa, on the other. It is also seen in the similarity of coastlines of the opposite continents, which suggested to A. Wegener the idea of continental drift, and in the similarity between the outlines of the outer edges of shelves and continental slopes on both sides of the ocean.

It must be emphasized that the above-mentioned peculiarities in the location and the strike of the basic forms of ocean relief, indicative of the symmetry relative to the Mid-Oceanic Ridge axis, at the same time show the existence of circumcontinental zonality in the structure of submarine relief. Successive replacement of geomorphological provinces (zones) from the continents to the Mid-Oceanic Ridge axis is clearly observed on the ocean floor. The symmetry and the circumcontinental zonality are thus mutually connected and complement each other. This is obviously indicative of the fact that the formation of ocean floor morphostructure has been caused by global processes and proceeded successively from the coasts of surrounding continents and on both sides of the Mid-Oceanic Ridge axis.

However, in the structure of the Atlantic Ocean floor relief there exist disturbances in the described symmetric scheme, introducing elements of asymmetry and serving as evidence of the heterogeneity of the Earth's crust. The largest of them are the Mexicano-Caribbean and the South Antillean Transition Zones situated on the western margin of the ocean and having no analogues on its eastern margin. The presence of these zones, including the basins of marginal seas, island arcs, and oceanic trenches, is a result of the effect of the Pacific Ocean Mobile Belt, on which the Atlantic segment of the Earth's crust borders here. These transition zones are evidently the elements introduced from the outside into the morphostructure of the Atlantic Ocean floor, and, therefore, they do not disprove its symmetry.

The less significant forms of ocean floor relief, causing its asymmetry, are individual submarine elevations, plateaus and ridges found on one of the sides of the Mid-Oceanic Ridge and having no analogues on the other. Such forms are rather numerous, but their relatively small size and limited occurrence can, in our opinion, only serve as indications of the manifestation of some local processes that resulted in their formation. These processes are, evidently, superimposed on the global processes. Disturbances in the symmetric scheme of the ocean floor relief structure are also produced by individual ridges, scarps, blocks, deeps, seamounts, and hills scattered in many places. All these disturbances, without disproving the general scheme, show the great diversity of relief-forming factors, both basic and secondary, creating all the variety of submarine relief forms.

The described symmetry and circumcontinental zonality manifest themselves in the Atlantic Ocean in the direction transverse to its axis. If, however, one examines the submarine relief structure in the longitudinal direction it is possible to note distinctly expressed morphological changes reflecting the non-uniformity of the Earth's crust structure, e.g., along the Mid-Atlantic Ridge and its rift zone [32, 64]. The structural heterogeneity manifests itself in the variation (along the strike) of the width, mean depth, and nature of dissection of the rift zone and the ridge as a whole, in the differences in size and the frequency of the occurrence of transverse faults, as well as in others factors. These variations in the structure of the submarine relief along the Mid-Oceanic Ridge are also manifested to a certain extent in the surrounding oceanic basins, and this is best seen in the width, changing from north to south, and the mean depth of basins, generally correlated with similar changes on the ridge. As far as the continental margins are concerned, the alterations in the character of submarine relief in the direction from north to south are caused by the influence of the specific features of the adjoining parts of land: their geological structure, history of development, occurrence of glaciations, climatic characteristics of coastline genesis, etc. This has resulted in the formation of dissected shelves of glacial regions, level shelves of middle latitudes, and the presence of coral structures on the tropical zone shelves.

Therefore, consideration of changes in the ocean floor relief structure taking place along its axis from the north to the south leads, firstly, to the conclusion that there exist nonuniformities in the development of global processes that have resulted in the formation of ocean floor morphostructure. These nonuniformities are, obviously, quite admissible, since our planet is not a homogeneous body. On the contrary, as shown by geological and geophysical data, the crust and the upper mantle of the Earth are characterized by a considerable structural nonuniformity. It must inevitably lead to dissimilar manifestation of global processes in different regions. Secondly, the characteristic features of submarine relief structure of the Mid-Oceanic Ridge and oceanic basins, on the one hand, and the continental margins, on the other, indicate the presence of diverse global processes underlying their formation. For the former the determining factor was, as we see it, the spreading of the ocean floor and for the latter – the subsidence of continental margins in the course of continental drift.

CHAPTER 2

The Structure of Sedimentary Series and its Role in the Morphostructure of the Ocean Floor

Sedimentation is known to be the most powerful exogenous relief-forming factor on the ocean floor. Accumulation of sedimentary sequences results in the filling in of depressions, the shrouding of irregularities in the bedrock relief, and, in the end, the decline of the ocean floor, provided, naturally, that the region is tectonically stable or generally submerged. The rise of the ocean floor, on the contrary, results in the erosion of deposits and the exposure of bedrocks. Therefore,. the investigation of sedimentary sequence reveals the tendency and the rates of vertical movements of oceanic regions, and the determination of the age of sediments makes it possible to date tectonic events that have led to the formation of the present-day morphostructural plan of the ocean floor.

1. THE GENERAL STRUCTURAL SCHEME OF SEDIMENTARY SERIES

Seismic data obtained by the reflection and refraction methods indicate that the sedimentary series of the ocean is characterized by a broad range of longitudinal seismic wave velocities: from 1.5 to 4.5 km s^{-1}. In the velocity spectrum a maximum with the values of 1.7–1.8 km s^{-1} is distinguished, indicating the predominance and widespread occurrence of loose sediments on the ocean floor [71]. Velocities of the same order have been determined by direct measurements in columns of bottom sediments taken by pipes and in the cores of the Deep-Sea Drilling Project (DSDP) [74, 165].

The sedimentary sequence on the ocean floor is usually subdivided according to changes of seismic velocities, and consequently to densities, into three basic layers: loose (1.5–2.0 km s^{-1}), semiconsolidated (2.0–3.0 km s^{-1}), and consolidated sediments, or sedimentary rocks (3.0–4.5 km s^{-1}). Examinations of DSDP cores have shown that loose sediments consist of terrigenous and biogenic aleurites and deep-water clays; semiconsolidatated sediments – of lithified aleurites, marls, volcanic ashes, and loose limestones; consolidated sediments – of solid limestones, flints, and lava sheets [165]. In sections of sedimentary sequences density and degree of lithification, as a rule, increase with depth, i.e., we observe a practically gradual transition from loose sediments to sedimentary rocks. This picture is disturbed by the presence of interlayers of relatively more dense rocks, forming a characteristic stratification recorded by the profilograph.

Investigations by the continuous seismic profiling method (CSPM) performed in the last 10–15 years have made it possible to identify the basic structural features of the sedimentary sequence on the floor of the ocean [9, 71, 121, 126–129, 137]. The thickness of sediments is minimal near the crest of the Mid-Atlantic Ridge and gradually increases on both sides towards the continental margins. At the foot of the continental slope there are, as a rule, foredeeps filled with very thick layers of sediments. In the section of a sedimentary sequence several reflecting horizons are singled out, of which the most distinct and consistent along the strike are horizons A and B. The horizon A where it is singled out in the CSPM records, invariably has a relatively flat, slightly undulating surface, not always correlating with the uneven relief of the underlying basement. This horizon is traced in the sedimentary thickness of oceanic basins, but in the zone of abyssal hills, as one approaches the flanks of the Mid-Atlantic Ridge, it is wedging out and merges with the horizon B. The latter, as shown by the data of the DSDP, represents a surface of the oceanic basement composed of basalts of different age, from the Neogene-Quaternary near the crest of the Mid-Atlantic Ridge to the Late Jurassic in the regions of the continental rise. The horizon A is mostly composed of interlayers of Eocene flints, but in some places it is represented by turbiditic interlayers or lithified sediments (aleurolites) dating from the Late Cretaceous to the Palaeocene. Reflecting boundaries above the A horizon are most commonly formed by turbidites, aleurolites or thin flint interlayers. Below the A horizon, reflecting boundaries are mainly represented by flint and limestone interlayers.

Loose sediments distinguished by seismic data are mainly deposited above the A horizon. In some places they are also traced below this horizon, mostly in the regions where it is of the Eocene age. Consequently, the sequence of loose sediments constituting the upper portion of the sedimentary section on the ocean floor is mostly represented by Cenozoic deposits, which is confirmed by the DSDP data [165].

Semiconsolidated and consolidated sediments are deposited in the lower part of the sedimentary section. The boundary between them varies considerably from one place to another. Besides that, consolidated sediments have a limited distribution and are only encountered near the continental margins and in some places in the Caribbean Sea. On the whole, judging by the DSDP data [165], the semiconsolidated and consolidated sediments are represented in the ocean by the Jurassic-Cretaceous, and in the Caribbean Sea by the Cretaceous and Lower Cenozoic sediments.

On continental margins, including the shelf and the continental slope, the structure of sedimentary thickness is different. Considerable differences in the structure and thickness of sediments are observed here, caused by the heterogeneity of the geological structure of continents. The technique of CSPM on continental margins, where consolidated sediments predominate, makes it possible to study only the upper portion of sedimentary section. The main data on its structure have been obtained by the refracted waves method and by drilling on the coast and the shelf. As shown by these works, non- or slightly dislocated sediments in continental margins lie on a heterogenous basement represented by an eroded surface of folded structures of varying age, from the Precambrian to the Late Palaeozoic, and made up of effusive and highly metamorphic rocks. This surface is rather distinctly recorded on seismic sections. The age of sedimentary rocks, occurring on it, is, as a rule, from the Triassic-Jurassic to Cenozoic, and the rocks are represented by an almost continuous section, with the exception of the regional interruptions in sedimentation. Packs of sedimentary rocks on the shelf of open coasts are usually monoclinal, inclined towards the ocean, the more ancient layers having a greater inclination than the overlying younger ones. This may by explained by the prolonged submergence of continental margins during the Meso-Cenozoic [208]. It is, probably, genetically connected with the formation of foredeeps at the foot of the continental slope.

The general picture of the distribution of sedimentary cover thicknesses in the Atlantic Ocean is shown on the schematic chart (Figure 12) compiled by us on the basis of the previously published charts and DSDP data [50,71, 126,137,165], as well as some supplementary data obtained in the cruises of R/V *Akademik Kurchatov*. Estimates of the volumes of sediments lying on the floor of the Atlantic Ocean have been performed and their results are listed in Table III.

TABLE III
The volume of sedimentary cover of the Atlantic Ocean

Geostructural provinces	Meso-Cenozoic		Cenozoic	
	$\times 10^6$ km^3	%	$\times 10^6$ km^3	%
Submarine continental margins	50	47	15	35
Foredeeps	12	11	4	9
Transitory zones	6	5	4	9
Oceanic basins	35	33	16	38
Mid-Atlantic Ridge	4	4	4	9
Total	107	100	43	100

As seen from Table III, the volume of Meso-Cenozoic deposits in the ocean reaches 107 million km^3, but their distribution within the geostructural provinces is extremely irregular. Their largest amount is concentrated

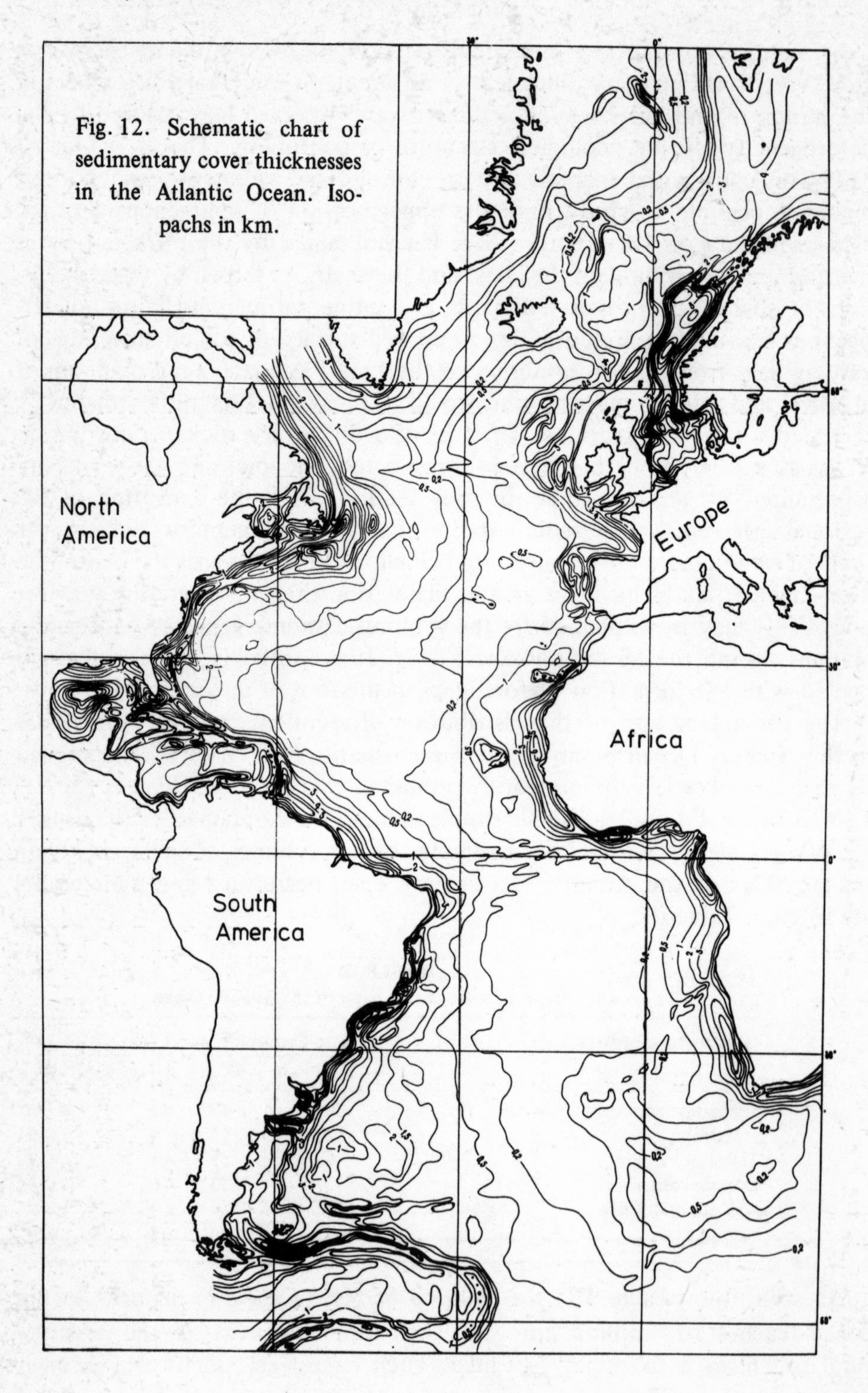

Fig. 12. Schematic chart of sedimentary cover thicknesses in the Atlantic Ocean. Isopachs in km.

on continental margins and in foredeeps, reaching 58% of the total volume, though the area of these provinces is not greater than 28% of the total area of the ocean. On the floor of the oceanic basins the volume of Meso-Cenozoic deposits is also appreciable. These are scattered over an area reaching 39% of the total area of the ocean. There are few sediments on the Mid-Atlantic Ridge and in the transition zones which are geologically young regions. A different picture is observed if one considers the Cenozoic deposits only. Their amount in oceanic basins is almost equal to that lying on continental margins and in foredeeps. Considerably greater is the share of sedimentary cover in the transition zones and on the Mid-Atlantic Ridge, where practically all sediments are of the Cenozoic age. This indicates, on the one hand, a considerable growth of sedimentation area at the end of the Mesozoic and in the Cenozoic, and, on the other hand, a higher intensity of Cenozoic sedimentation on the bottom of oceanic basins, which is close to the sedimentation intensity on continental margins.

2. SEDIMENTARY COVER OF THE CONTINENTAL MARGINS AND FOREDEEPS

The sedimentary cover on the shelf of Greenland is on the whole not very thick, gradually increasing to the edge of the shelf approximately up to 1 km and on the continental slope decreasing again. To the south of the Dutch Strait, in the zone where Precambrian structures outcrop on the shelf, sedimentary thickness seems to be even less, in the range of tens and hundreds of metres, and relatively greater thicknesses are associated with local depressions in the basement. On the continental slope the bassment subsides abruptly along a system of faults. The sediments cover it with a thin, intermittent layer, but at the foot of the slope the thickness of sedimentary cover is seen to increase considerably owing to the formation of accumulative series. The thickness of sediments in the foredeep here reaches 2–3 km [125, 178].

Near the coasts of Labrador and Newfoundland the basement composed of Precambrian and Caledonian folded structures is practically exposed in the zone of the coastal shoal, but along the line of longitudinal trenches it goes sharply down under the outer part of the shelf. To the east of Newfoundland the sedimentary sequence at the outer edge of the shelf is more than 3 km thick, in some places up to 8 km. It fills up the foredeep which, thus, also merges with the outer part of the shelf [215]. Borehole 111, drilled on the Orphan Hill in the lower portion of the continental slope, encountered Upper Cretaceous sediments at a depth of 250 m. In the basement of the sedimentary section here obviously even more ancient rocks are deposited, possibly Jurassic. At the same time, on the Flemish Cap is located a prominence of the basement covered only by a thin layer of sediments [164].

The coastal plain and the shelf of the U.S.A. from Nova Scotia to Florida are made up of monoclinal sedimentary rocks of the platform mantle inclined towards the ocean. The folded Hercynian basement forms a flexure under the shelf, where the thickness of sediments ranges from 3 to 6 km. At the outer edge of the shelf the basement surface rises in the form of a crest up to 1–3 km high, after which it sharply submerges towards the foredeep where the sedimentary thickness reaches 6–9 km [127]. Sedimentary cover on the shelf is represented by an almost uninterrupted section from Upper Jurassic to Tertiary deposits, which is indicative of a prolonged period of continental margin subsidence (Figure 13). On steep

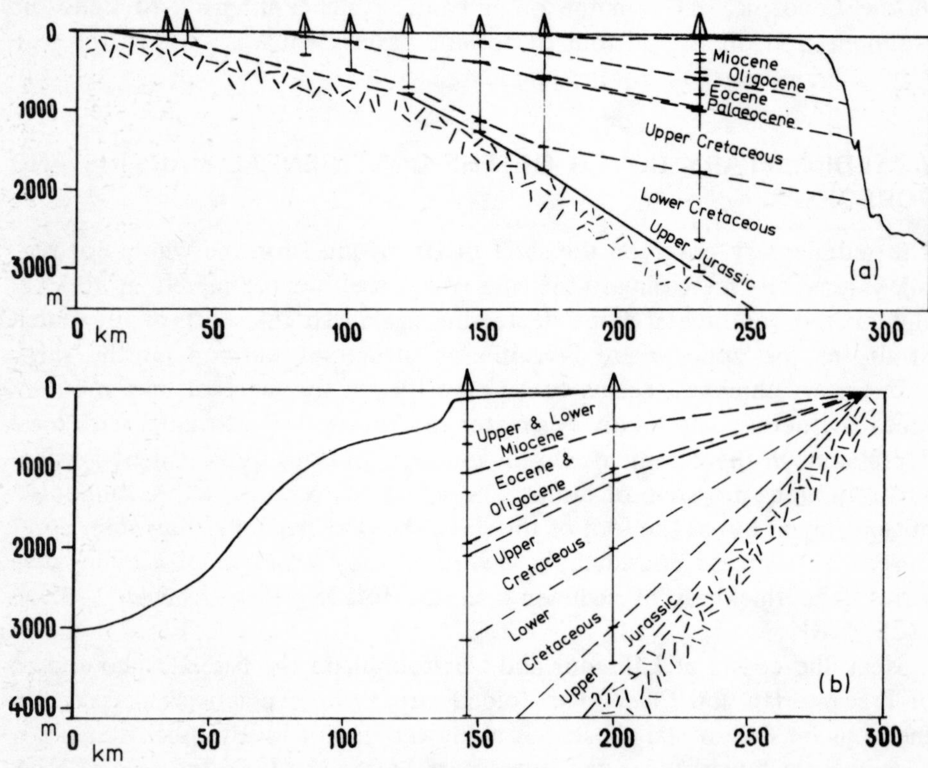

Fig. 13. **Sedimentary sequences in the continental margins of North America (a) and Africa (b) according to drilling data [208].**

scarps of the continental slope and on the walls of canyons the sediment layers outcrop to form structural terraces, and on gentle sections they bend downward in a flexure-like manner to the foot of the slope. Neogene-Quaternary and present-day sediments on the U.S.A. shelf are deposited in an interrupted fashion as separate spots, which indicates the presence of interruptions in sedimentation associated with the fluctuations of the ocean level at the end of the Neogene and during the Pleistocenous glaciations.

A considerable submergence of the continental basement is observed in the regions of Florida, the Blake Plateau, and the Bahama Platform [215]. The thickness of sedimentary cover on the Blake Plateau reaches more than 6 km, but at the outer edge it diminishes considerably as a result of the rise of the basement in the form of a boundary crest. Boreholes 389, 390 and 392, drilled at the edge of the plateau under eroded Palaeogene sediments, encountered Cretaceous deposits. On the Bahama Platform the sedimentary cover with a thickness equal to 4–6 km is mainly composed of coral limestones from the Upper Jurassic to Recent, known to be formed on shoals. This is indicative of the extent of the submergence of the platform during the Meso-Cenozoic.

Around the Gulf of Mexico the basement is composed of Hercynian folded structures, subsiding everywhere towards the basin. The largest thicknesses of sedimentary cover are observed in the northern part of the gulf. A deep basement flexure filled with sediments up to 12–15 km thick extends here along the coast of the U.S.A. In the vertical section of sedimentary sequence one can see a regular alternation of formations from lagoon to marine types, which is indicative of the gradual submergence of the continental margin. On the Campeche Bank the thickness of the sedimentary cover is equal to 1–2 km. It is composed of limestones dating from the Cretaceous to Recent. In the Sigsbee Basin occupying the central part of the Gulf of Mexico the thickness of sedimentary cover reaches 8–10 km [133]. Judging by seismic data, the folded-metamorphic basement is wedging out here and is replaced by the 'basalt' layer rocks. Deep-sea drilling has disclosed terrigenous-carbonate deposits from the middle Miocene to Recent. A large number of turbiditic interlayers is observed which indicates the major role of suspension currents in the formation of sedimentary cover. Diapir structures are widespread, represented in the present-day relief in the form of rounded knolls. In borehole 2, drilled on one of the Sigsbee Knolls under the sediments lying at a depth of 144 m, were found calcite, gypsum and manifestations of gas and oil, which confirms the opinion that these knolls are of salt-dome origin [165].

On the shelf of South America along the coasts of Guiana and Brazil, the structure of sedimentary cover is determined by projections of Precambrian shields (the Guianian and the Brazilian), whose marginal parts are diversified by a series of flexures [101]. Here on the eroded surface of the basement lie sediments dating from the Upper-Jurassic to Quaternary with some interruptions in their section, indicative of regressions and transgressions of the ocean. Lagoon sediments prevail almost everywhere at the base of the section, with marine shelf sediments lying above them. Cenozoic deposits near the coast form about a third of the sedimentary section, on the outer part of the shelf reaching half of the section. Sediment layers are inclined towards the ocean. Their total thickness on the shelf reaches: near Recife – 0.5 km, to the south of Recife – 3 km, on the Abrolhos Bank – 3.5 km, in the Amazon Foredeep more than 4 km. On the outer edge of the shelf and on the continental slope near the mouths of the Amazon, and on the marginal Amazon Plateau, a thick submarine cone is located, composed of sediments dating from the Miocene to Recent, superimposed on the more ancient deposits. In foredeeps belting the continental margin of Brazil the thickness of sedimentary cover ranges from 2 to 3–4 km.

Along the Argentina coast the eroded surface of the Palaeozoic folded basement on the coastal plain and on the shelf is gradually subsiding towards the ocean. At the outer edge of the shelf this surface forms a wide crest whose outer slope steeply descends to the foot of the continental slope [188]. The basement here is diversified by transverse flexures with the thickness of sediments reaching 5–6 km, while on the remaining territory of the shelf it amounts to 1–2 km. The sedimentary cover is represented by a section of deposits from the Upper Jurassic to the Neogene-Quaternary age. By the variation of facies from the lagoon to the shelf types one can witness the successive submergence of the continental margin in the Meso-Cenozoic. In the foredeep extending in the form of a wide belt along the foot of the continental slope the thickness of sediments reaches 3–4 km.

To the south of the Falkland Islands lies another large transverse flexure, formed by a basement projection and extending from the shelf in latitudinal direction along the trench separating the Falkland Plateau from the Scotia Ridge, with sediment thickness in it reaching 7–9 km. Along the northern and southern edges of the Falkland Plateau basement projections are traced in the form of crests, between which extends a longitudinal flexure filled up by sediments of more than 3 km thickness [134]. In borehole 330, drilled in the eastern part of the plateau at the depth of 556 m below the floor level, Precambrian granites and gneisses have been found, with Mesosoic sediments discordantly bedding over them.

Along the eastern margin of the Norwegian-Greenland Basin almost everywhere is traced a deep flexure of the outer part of the shelf filled by a thick wedge of sediments. On the continental slope the sediments have a monoclinal bedding inclined towards the foredeep where their thickness reaches more than 3 km [116]. On the shelf of Norway the surface of folded basement is practically exposed in the strandflat zone, but on the outer part of the shelf it sharply descends along a system of edge faults. Along the foot of the continental slope and in the zone where the Norwegian Plateau adjoins the shelf, there is a wide foredeep, filled with sediments whose thickness is up to 6–8 km [126]. In the outer part of the Norwegian Plateau the thickness of sedimentary cover is sharply reduced because of the basement projection forming a kind of structural nucleus separated from the shelf by the foredeep. In borehole 338, drilled in this area at the depth of 437 m below the floor level, basalt layers have been found under the Lower Miocene terrigenous deposits.

In the northern part of the North Sea the Caledonian folded basement lies at relatively shallow depth, but in the central and southern parts of the sea the Hercynian folded structures are submerged to a considerable depth, forming a spacious flexure filled with Upper Palaeozoic and Meso-Cenozoic deposits whose total thickness is up to 12 km [45]. On the floor of the sea the relict Quaternary sediments prevail, represented by derived and reworked moraine deposits left by Pleistocene glaciers.

On the shelf, along the western coasts of Scotland and Ireland, there is a gradual subsidence of the folded basement surface from the coasts to the edge of the shelf and a sharper one, along a system of faults, on the continental slope, reaching its maximum in the Faeroe-Shetland and Ireland Trenches. The foredeeps separating the blocks of the Rockall Plateau and of the Iceland-Faeroe Rise from the continental margin extend here. The thickness of sedimentary cover at the outer edge of the shelf reaches 0.5 km and in the foredeeps up to 3 km [137].

On the Rockall Plateau the basement surface is raised on the Hutton and Rockall Banks and flexured between them (Figure 14), where sedimentary thickness reaches more than 1.5 km [207]. Borehole 116 drilled in the flexure disclosed a section of carbonate silts dating from Pleistocene to Oligocene and below passing into Eocene limestones. In borehole 117, drilled on the slope of the Rockall Bank at the depth 310 m below the floor level, volcanic conglomerate has been found under Palaeocene clays. Small sedimentary thicknesses are also observed on the Faeroe shelf and on the Iceland-Faeroes Rise. In deep–sea boreholes numbers 336, 337, 352 basic lavas have been encountered on the slopes of the Rise under Eocene deposits. The sedimentary section is represented by terrigenous-volcanic sediments with a clearly defined interruption in the Miocene, indicating the existence of subaerial conditions here at that time.

To the south of Ireland and along the coast of France the Hercynian folded basement gently descends towards the shelf edge and more steeply to the foot of the continental slope. The sedimentary thickness on the outer part of the shelf amounts to about 2 km. On the continental slope it decreases as a result of erosion, but at the foot of the slope in the foredeep it increases up to 3–4 km. To the south of Ireland and in the English Channel transverse depressions in the basement have been found. These are filled with Upper Palaeozoic and partly Mesozoic deposits more than 3 km thick.

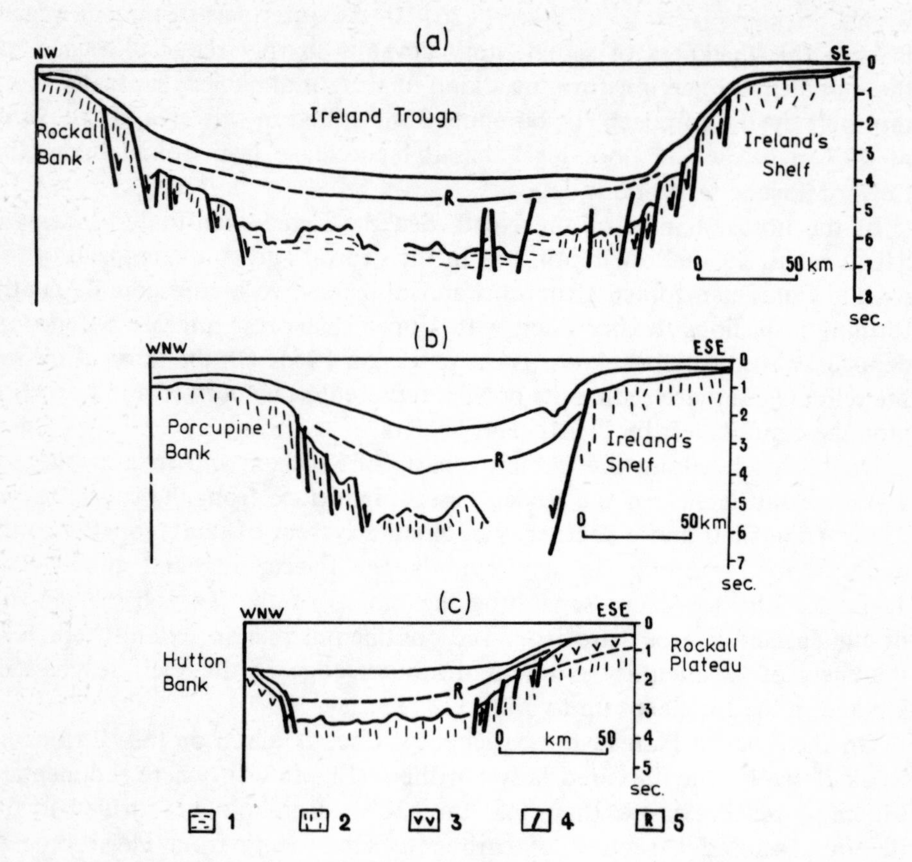

Fig. 14. Structure of sedimentary sequences and basements in the regions of Ireland Trough (a), Porcupine Bank (b), and Rockall Plateau (c), according to CSP data [207]. (1) oceanic basement; (2) continental basement; (3) Paleocene volcanites; (4) faults; (5) Eocene boundary in sediments.

At the shelf edge Palaeozoic deposits are wedging out and the sedimentary section is represented only by Meso-Cenozoic sediments [201].

Along the coasts of the Iberian Peninsula a stepwise submergence of the folded basement is observed along a system of faults bounding the continental margin. The sedimentary cover has a variable thickness, in individual depressions (e.g., in the eastern part of the Iberian Plateau) reaching more than 1 km. On gentle parts of the slope the sediments lie in the form of inclined layers whose thickness increases downwards, and on steep scarps they are crumpled and disturbed by slumps [225]. In the foredeep to the south of the Iberian Plateau the thickness of sedimentary cover is more than 2 km. Borehole 298 drilled in this area passed through 1700 m of terrigenous-carbonate deposits, from the Quaternary to the Lower Jurassic, and still did not reach the basement.

At the north-western coast of Africa the Hercynian folded basement of the Haut Atlas mountains extends on the adjoining shelf, on the whole lying not very deeply, but on the continental slope it descends sharply along scarps to the foredeep where the thickness of sediments reaches more than 2 km. Further to the south, along the western coast of Africa, the structure of sedimentary cover of the continental slope is determined by alternating projections and flexures of the Precambrian folded basement. In the regions of basement projections the sedimentary cover is very thin near the coast, but towards the shelf edge its thickness increases up to 0.5 km. In flexures, the largest of which is the Senegal Syncline, the basement surface gently submerges towards the foot of the continental slope. A sequence of Meso-Cenozoic deposits is bedded on it, reaching a total depth of 3–4 km, of which Cenozoic deposits account for no more than 1 km [225]. On the continental slope this sedimentary thickness is either cut by the erosive surface or decreases in thickness, but in the foredeep it increases again up to 2–3 km. The facies of the sedimentary section, according to the data of a borehole in the area of Cap Blanc, changes upwards from the continental to the coastal type (see Figure 13), which is indicative of the subsidence of the continental margin starting from the Lower Jurassic [208]. In the foredeep, according to DSDP data (boreholes 368, 369, 197), there is a continuous section of terrigenous-carbonate deposits with turbidities and flint interlayers from the Quaternary to the Lower Jurassic.

Along the northern coast of the Gulf of Guinea is traced a series of small marginal flexures separated from the land by faults. The flexures alternate with projections of the basement, where the sedimentary cover is not thick, whereas in the flexures its thickness reaches 5–6 km and in the region of the mouths of the River Niger up to 8 km [189]. Cenozoic deposits form about a third of the sedimentary section. These flexures, merging at the foot of the slope, form a single foredeep.

A similar picture is observed along the coasts of Gabon, Congo, and Angola [129, 143]. Here there is also a series of marginal flexures, filled with sediments from the Cretaceous to the Neogene-Quaternary with a thickness up to 3–4 km. Their facies change upwards from the continental to the lagoon and shallow marine types. In the Cenozoic deposits there are a series of interruptions, the largest one associated with the Oligocene. In the foredeep the total thickness of sedimentary cover reaches 2–3 km. This flexure also extends between the shelf and the Walvis Ridge separating the latter from the continental margin [146].

At the south-western coast of Africa there are no distinctly expressed marginal flexures. The folded basement here gradually submerges from the coast to the foot of the continental slope [191]. The sedimentary thickness gradually increases in this direction and in the foredeep reaches 4 km. The main part of the section on the shelf is composed of Cretaceous deposits, with Cenozoic sediments up to 0.5 km thick discordantly bedding on their eroded surface. In the foredeep the thickness of Cenozoic sediments (according to the data of borehole 360) increases up to 1 km.

3. THE SEDIMENTARY COVER OF TRANSITION ZONES

The structure of the sedimentary cover of the Caribbean and the South Antillean Transition Zones have their specific features. The basement here is formed by the surface of Meso-Cenozoic folded structures on the submarine slopes of sub-latitudinal branches of the Antillean and the Scotia Ridges and by the surface of volcanogenic rocks bedding on the floor of the basins and slopes of volcanic island arcs. The data of deep-sea drilling indicate that throughout the basins the basement surface is represented by basalts. The overlying sedimentary cover is composed of Upper Cretaceous semiconsolidated deposits (mainly limestones) and Cenozoic loose sediments, the latter composing the larger part of the section. The boundary between them most often, though not always, coincides with acoustic horizon A formed by flint interlayers of Middle Eocene age [165].

The sedimentary cover in the Caribbean Sea at the crest of the Antillean Ridge (between islands) and its slopes is, on the whole, not very thick, and on steep parts it is interrupted by the outcrops of rocks of folded or volcanogenous basement. At the foot of the slopes, however, are accumulative series which subsequently pass into the cover of the basin floor. Along the continental margin, on the contrary, the structure of sedimentary cover is more complex. Except for individual projections of folded basement, where the thickness of sediments is not great, most of the shelf is formed by marginal flexures made up of a sedimentary cover up to 3–5 km thick. These flexures are especially noticeable along the coast of Venezuela [133].

On the ridges and rises in the inner part of the Caribbean Sea the sediments are also distributed irregularly. The crest and the upper parts of the slopes of the Cayman Ridge are covered by a thin and interrupted layer of Neogene-Quaternary sediments. At the foot of the slopes, however, the thickness of sediments increases to several hundred metres. On the Nicaragua Rise, which is a more ancient structure, the thickness of sedimentary cover is much greater – up to 2 km, though on steep scarps it decreases noticeably as a result of slumps. Borehole 152 on the south-eastern slope of the rise disclosed a basalt basement under loose Eocene-Quaternary sediments and semiconsolidated Cretaceous-Palaeogene deposits with a total thickness of 470 m. A comparatively think sedimentary cover is found on the Beata Ridge. For example, borehole 151 drilled on the southern extremity of the ridge passed through carbonate deposits dating from the Pleistocene to the Upper Cretaceous reached the basalt basement at the depth of 380 m below the sea floor

(Figure 15). A thicker sedimentary cover (more than 1 km) lies on a longitudinal flexure on the Aves Rise, while on the marginal rises of the basement its thickness sharply decreases [119].

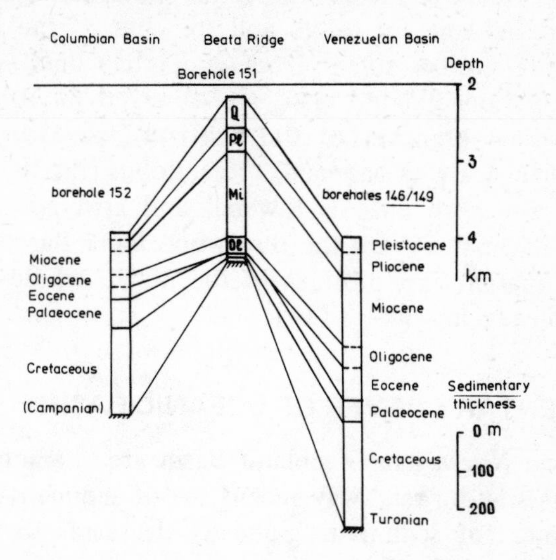

Fig. 15. Sedimentary sequences of the Caribbean Sea according to deep-sea drilling data.

The thickness of sediments is considerable, though unequal, in the basins of the Caribbean Sea. In the Yucatan and the Colombian Basins sediments of more than 2 km thickness have accumulated, while in the Venezuelan Basin they are not more than 1 km thick. Near the continental margin the thickness of sediments markedly increases owing to the formation of accumulative series. In DSDP boreholes in the Venezuelan and the Colombian Basins the sedimentary section is represented by loose Neogene-Quaternary sediments and semiconsolidated Palaeogene-Cretaceous deposits, largely of carbonate composition. A fairly thick sedimentary cover (up to 2–3 km) is observed in the basins on both sides of the volcanic arc of the Lesser Antilles. Even greater thicknesses of sedimentary cover are found in the marginal flexure on the eastern slope of the Barbados Ridge – up to 6 km, which resulted here in the inversion of submarine relief from the trench to the edge swell [133].

A layer of sediments up to 1–2 km thick has been accumulated in the deep-water Puerto Rico and Cayman Trenches serving as traps for sediments. But this does not compensate for the flexure of the Earth's crust, as these trenches are tectonically active structures, their formation continuing until now [118].

A similar picture of the sedimentary thickness structure is observed in the South-Antillean Transition Zone. In the eastern part of the Scotia

Sea the thickness of sediments represented by loose and semiconsolidated deposits is about 2 km. Near the foot of the Scotia Ridge and in the marginal trenches bounding the basin the sedimentary thickness increases up to 3 km. In the western part of the Scotia Sea the sedimentary cover is markedly reduced, and in the zone of ridges and trenches of the Drake Passage it becomes interrupted and in some places completely disappears. Quite thick sediments, mostly consolidated ones, are observed on the lowered blocks of the sublatitudinal branches of the Scotia Ridge. On the block of the South Orkney Islands, e.g., a basement projection is traced along its northern edge, while the southern edge is lowered and covered by a sedimentary sequence up to 2 km thick. On the Burdwood Bank the basement is raised on its southern margin and subsides to the north, where the thickness of sedimentary cover reaches 6 km [134].

4. THE SEDIMENTARY COVER OF OCEANIC BASINS

The basins of the Norwegian-Greenland Basin are characterized by a considerable thickness and a relatively recent age of sediments. In the Lofoten Basin the thickness of sediments gradually decreases to the west from 2 to 1 km. In the Norwegian Basin the thickness varies because of the uneven relief of the basement, whose largest uplifts rise above the sediments in the form of seamounts of the Aegir Ridge. In the troughs between the seamounts the sedimentary thickness reaches more than 1.5 km [126]. According to the DSDP data (boreholes 343 and 345), the sedimentary section is mainly represented by loose terrigenous deposits, the most ancient of them having an Eocene age.

The Irminger and the Labrador Basins are characterized by a greater thickness of sediments, reaching 2–3 km. Near the Reykjanes Ridge and over the Mid-Labrador Ridge, however, the sedimentary thickness does not exceed 1 km and in some places (on mountain slopes) is reduced to a minimum [73, 125]. In borehole 112 drilled in the south-eastern part of the Labrador Basin the sedimentary section is represented by loose terrigenous and terrigenous-flint deposits up to 662 m thick, the oldest of them having early Eocene age, with a basalt surface disclosed below them. Borehole 113, drilled in a deep of the crest portion of the Mid-Labrador Ridge, passed through loose sediments dating from Pleistocene to Miocene with a thickness of more than 920 m, almost reaching the basement surface.

In the Newfoundland Basin the sediments are less thick. They are clearly seen to increase from the border of the Mid-Atlantic Ridge to the foredeep from 0.6 to 1.5 km. A similar picture, only more markedly expressed, is observed in the North-Western Atlantic Basin [137]. On the Sohm and the Hatteras Abyssal Plains the thickness of sedimentary cover reaches 2 km, on the Nares Abyssal Plain – about 1 km, and in the zone of abyssal hills it decreases to 0.3–0.5 km. Not only the sedimentary thickness but also the age of basal deposits increase towards the foredeep. According to the DSDP data, in the vicinity of the Mid-Atlantic Ridge the oldest sediments are of Late Cretaceous age (Maestrichtian-Campanian), whereas in the western and southern parts of the Hatteras Abyssal Plain they are Late Jurassic (Tithonian-Oxfordian). It is necessary to note that on abyssal plains more than 5000 m deep the upper part of sedimentary section (Neogene-Quaternary deposits) is, as a rule, represented by almost noncarbonate or weakly carbonate silts, while the lower part of the section is composed exclusively of carbonate deposits [165]. This may indicate that previously the basin floor was situated above the critical depth of carbonate dissolution and then subsided.

The Blake-Bahama and the Newfoundland Ridges are composed of bodies of sedimentary material up to 2–3 km thick. This is exemplified by the fairly well studied Blake-Bahama Ridge where it is seen that there is a slight elevation of the basement surface under it, this rise having, evidently, served as a primary barrier over which the accumulative ridge started to form under the effect of hydrodynamic factors [190]. Several boreholes of DSDP (boreholes 102–104) passed here through the upper sequence of loose sediments dating from Holocene to Miocene and reached at the depth of 620 m a strongly reflecting horizon represented by the roof of semiconsolidated deposits.

On the Bermuda Rise sediments have a variable thickness, increasing up to 1 km in basement depressions and decreasing on its elevations. The sedimentary section is mainly represented by carbonate silts with rare turbidite interlayers. The ages of the most ancient deposits are Late (borehole 9) and Early Cretaceous (borehole 386). Below them lies the surface of the basalt basement (Figure 16). The sedimentary cover on the Antilles Outer Swell has similar structure and age, only the total thickness of deposits is smaller here, amounting to about 0.5–0.6 km [118].

In the Guiana Basin the thickness of sedimentary sequence gradually increases from the Mid-Atlantic Ridge towards the foredeep from 0.5 to 2.5 km. In borehole 27 drilled in the northern part of the basin loose and semiconsolidated deposits have been found, those on top almost noncarbonate, those below carbonate. At the depth of 475 m from the floor level Upper Mesozoic clays have been found with aleurite interlayers, but the basement was not reached.

The Brazilian and the Argentine Basins are similar in the structure of their sedimentary covers, but the thickness in the former is on the whole less than in the latter. The thickness of sediments is everywhere observed to increase from the Mid-Atlantic Ridge towards the foredeeps: in the Brazilian

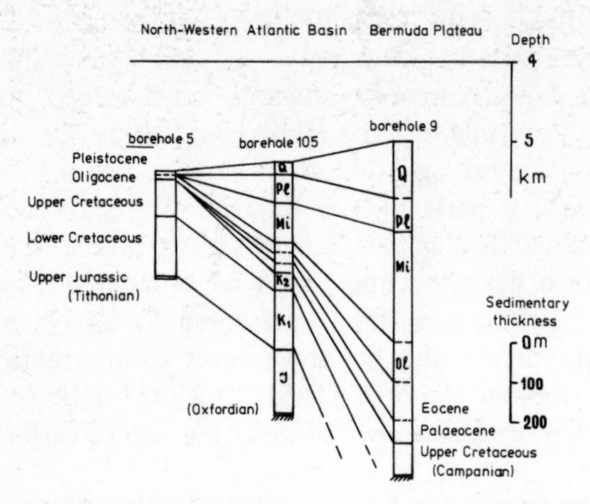

Fig. 16. Sedimentary sequences of the North-Western Atlantic Basin according to deep-sea drilling data.

Basin from 0.2 to 1.5 km, and in the Argentine Basin from 0.5 to 2 km and more. Local uplifts of the basement in the forms of ridges and small mountain ranges are found in many places under the sedimentary cover, and this results in sharp changes in the sedimentary thickness [137]. Boreholes 19 and 20 in the zone of abyssal hills of the Brazilian Basin disclosed the following section: from Pleistocene to Oligocene – red deep-water clays, from Eocene to Upper Cretaceous – carbonate silts and clays, the basement – basalt layers and breccias.

On the Rio Grande Plateau, as well as on the Seara Rise and the South Antilles Outer Swell, the sedimentary cover is generally not thick (0.2–0.5 km, in some places up to 0.8–1.0 km) and is uneven due to the irregularities of the volcanogenous basement. Two boreholes drilles on the Rio Grande Plateau disclosed carbonate deposits dating from Pleistocene to Upper Cretaceous (Maestrichtian), but did not reach the basement (Figure 17). Borehole 354 drilled on the northern slope of the Seara Rise encountered under the Pleistocene turbidite carbonate deposits, dating from Pliocene to Upper Cretaceous, with a total thickness of 900 m, and penetrated into the basalt basement [165].

In the Iceland Basin the sediments are of a great thickness (up to 1.0–1.5 km) and comparatively recent age, not going beyond the Cenozoic. Borehole 115 drilled almost in the centre of the basin disclosed only Pleistocene terrigenous deposits of more than 220 m thickness with numerous turbidite interlayers composed of volcanic ash.

In the North-Eastern Atlantic Basin there is a considerable thickness of sediments near the foredeep and on the Biscay Abyssal Plain (up to 2 km) and a rather fast reduction of it in the zone of abyssal hills (up to

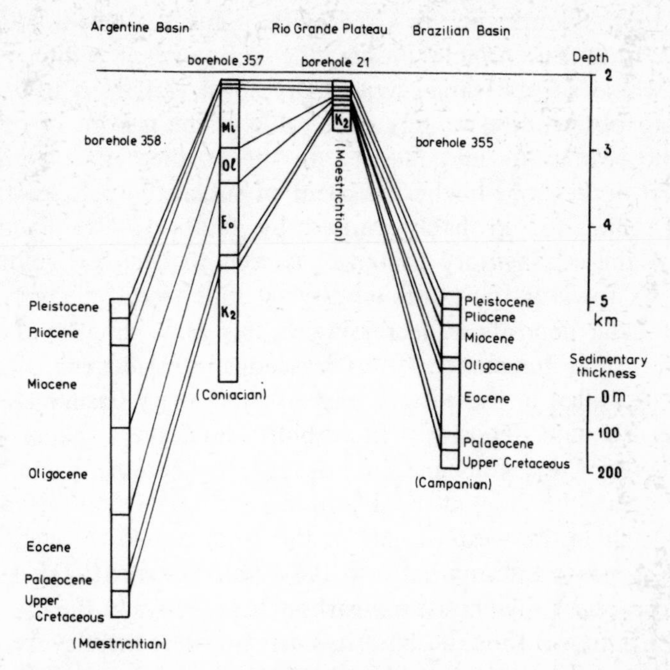

Fig. 17. Sedimentary sequences of Rio Grande Plateau and Brazil and Argentine Basins according to deep-sea drilling data.

0.5 km). Borehole 118 drilled close to this zone passed through 670 m of Miocene-Quaternary terrigenous-carbonate deposits with a large number of turbidite interlayers and at the depth of 750 m reached a basalt sill covered by Eocene carbonate clays.

A complex structure of sedimentary cover is observed in the region of the Iberian Basin and the Azores-Cape St Vincent Ridge. The presence of numerous seamounts and uplifts caused by the unevenness of the oceanic basement results in sharp fluctuations of sedimentary thickness. In depressions of the basement the sedimentary thickness is equal to 1–3 km, but it is sharply reduced and in some places interrupted on the slopes of the uplifts.

Borehole 120 drilled on the slope of the Gorringe seamount passed through 250 m of carbonate deposits, the most ancient of them having an Early Cretaceous age, and entered into the volcanogenous rocks of the basement. To the north of Madeira borehole 136 passed through 310 m of loose and semiconsolidated carbonate deposits dating from Pleistocene to Upper Cretaceous and reached the basement made up of tholeiite basalts. These data are indicative of a Cretaceous age of the oceanic basement surface and the seamounts in the eastern part of the Azores-Cape St Vincent Ridge.

Further to the south, in the Canary, the Cape Verde, the Sierra Leone, the Guinea, and the Angola Basins, the structure of sedimentary cover is on the whole of the same type. Only local variations of sedimentary thickness are observed, especially noticeable in the regions of oceanic rises and volcanic seamounts and massifs. In the northern part of the Canary Basin several depressions in the basement of sublatitudinal strike are found under the sediments, probably caused by faults [149]. Everywhere in these basins the sedimentary thickness is seen to decrease gradually from the foredeeps towards the zones of abyssal hills from 1.5 to 0.2–0.5 km. The age of basal deposits is decreasing in this same direction, from Early Cretaceous – later Jurassic to Late Cretaceous – Palaeocene. For instance, borehole 140 drilled in the eastern part of the Canary Basin passed through 650 m of carbonate deposits with turbidite interlayers dating from Pleistocene to Upper Cretaceous and did not reach the basement, that, according to the CSPM data, lies here at the depth of more than 1.5 km. Boreholes 137 and 138 drilled in the western part of the basin passed through more than 400 m of sediments and entered into the basalt basement. The sedimentary section is represented by almost noncarbonate red clays in the upper portion, then by terrigenous-carbonate deposits with turbidite interlayers, and below – by carbonate semiconsolidated Upper Cretaceous deposits.

The sedimentary cover decreases sharply on the slopes of elevations and volcanic seamounts. However, on the upper surface of plateau-like elevations the thickness of sedimentary cover increases again (Figure 18). For instance, on the Sierra Leone Rise the sedimentary thickness is equal to 0.8–1.0 km. Borehole 336 drilled at the eastern edge of the rise, passed through 850 m of carbonate deposits dating from Quaternary to Upper Cretaceous, but did not reach the basement. The block-type Walvis Ridge is also covered by sediments with a thickness reaching several hundred metres, and in local depressions of the basement – up to 1.5 km. According to the data, obtained by examining sediment cores [136] and from the data of dredging performed by the R/V *Akademik Kurchatov*, it can be assumed that the lower part of sedimentary section here is composed of semiconsolidated Palaeogene and Upper Cretaceous deposits, mainly foramineferal limestones.

The structure of sedimentary cover in the Cape and the Agulhas Basins is more complex and resembles the structure of the Argentine Basin sediments. Rather sharp changes in the sedimentary thickness are also observed here, caused by the presence of irregularities in the basement surface in the form of low ridges and local depressions. In the latter the thickness of sedimentary cover reaches 0.8–1.0 km, over basement uplifts it decreases to 0.2–0.4 km [137].

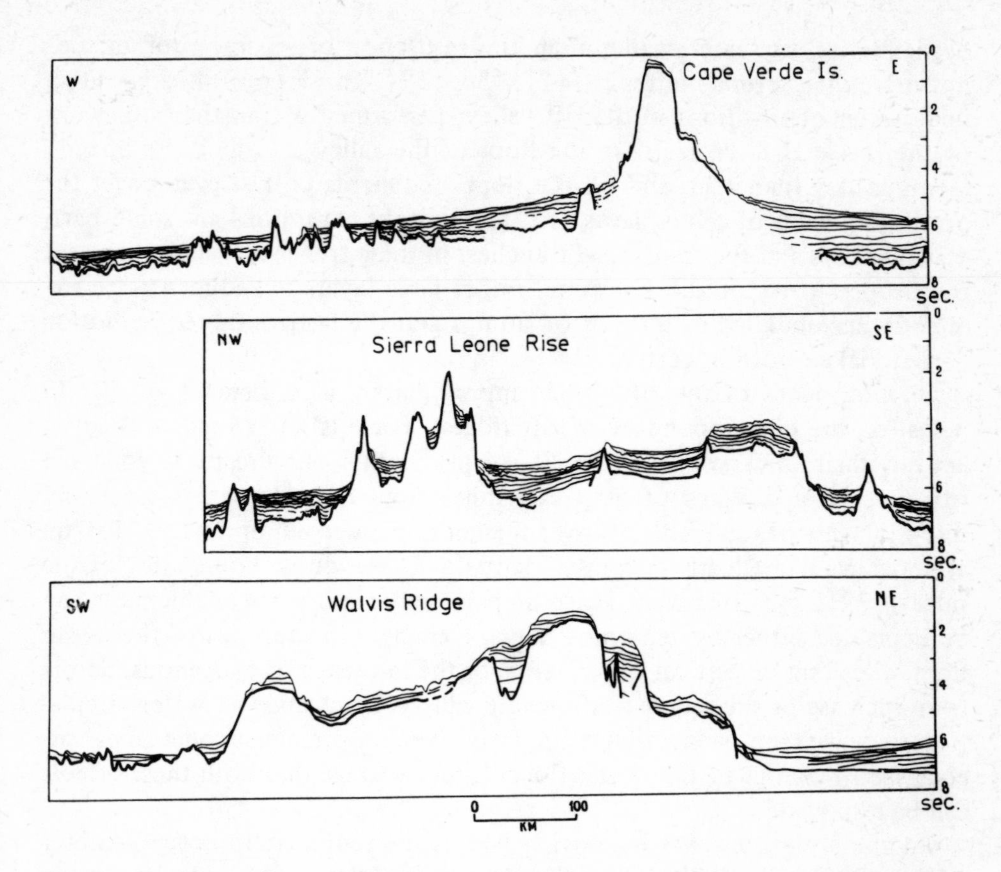

Fig. 18. Continuous seismic profiles of Cape Verde and Sierra Leone Rises and Walvis Ridge [225].

5. THE SEDIMENTARY COVER OF THE MID-ATLANTIC RIDGE

The whole zone of the Mid-Oceanic Ridge is characterized by a cluster-like distribution of sedimentary cover in the form of a kind of 'pockets' in the inter-ridge valleys, as well as by an increase in thickness on both sides of the rift valley. The age of sediments at the bottom of the sedimentary cover on the surface of the oceanic basement also increases in the same direction. On the slopes and tops of ridges and seamounts the sedimentary cover is frequently interrupted, exposing the surface of bedrocks. Sediments of the Mid-Oceanic Ridge are distinguished by their acoustic transparence; there are few reflecting surfaces in them. They mostly have carbonate composition, which shows the predominant role of biogenic sedimentation and the location of the ridge above the level of the critical depth of carbonate dissolution during the whole period of their formation.

Examination of a large number of CSPM profiles has shown that in the rift zone the sediments are either completely absent or their thickness is negligible, at any rate smaller than the resolution of seismic profiler, i.e., not more than several metres [9, 121, 136, 137]. This is confirmed by direct underwater observations in the rift valley, performed within the framework of the 'FAMOUS' Project. On the floor of the valley the thickness of sediments is less than 1 m, and on the slopes sediments do not even cover the protruding parts of pillow lavas [159]. The only exceptions are some parts of rift valleys and the transversal trenches. In these trenches, which represent the deepest parts of the rift zone, rather large layers of sediments are frequently accumulated as a result of slumps and the horizontal redistribution of material by bottom currents.

On both sides of the rift valley approximately at a distance of 20–30 miles (on the outer boundary of rift ridges) 'pockets' of sediments begin to appear, their thickness equal to 30–40 m. Further on, already beyond the rift zone, at a distance of 100–200 miles from the ridge axis a relatively sharp increase of sedimentary cover thickness is observed, up to 100–150 m, after which it gradually increases right up to the outer boundaries of the ridge [137]. This relatively sharp step in sedimentary cover thickness can be explained either by tectonic effects – changes in the rates of the ocean floor spreading or vertical movements, or the influence of exogenous factors – an increase in the sedimentation rate caused by changes in water circulation and plankton productivity, the latter being the main supplier of calcareous shell material to the ocean floor. It seems to us, that both these effects can be responsible.

On the Mohns and the Knipovich Ridges the sedimentary cover is rather thick and fills up practically all the inter-ridge valleys, and it also occurs on the tops of some seamounts [51, 126]. This agrees with the increased thickness of sediments in the depressions of the Norwegian-Greenland Basin and is, evidently, explained by its relatively small size and proximity to drift zones. At the same time, on the Iceland Ridge the sedimentary cover is practically absent or has a negligible thickness in inter-ridge valleys. This can probably be explained by the recent age of the Iceland Ridge, comparable to the age of the rift zone of Iceland. On the neighbouring Iceland Plateau the age of sediments is more ancient, and their thickness reaches 0.4–0.6 km. Borehole 348 drilled near the Iceland Ridge passed through a thickness of terrigenous sediments dating from Pleistocene to Upper Oligocene and at the depth of 544 m reached the basalt basement. Borehole 350 drilled at the eastern edge of the plateau disclosed a similar sedimentary section and also reached the basement, but the age of basal deposits here is Late Eocene.

Relatively increased thicknesses of sedimentary cover are observed on the Reykjanes Ridge [221]. Only in the narrow crest zone are sediments either absent or negligible. In borehole 409 drilled here, below Pliocene-Quaternary deposits 80 m thick was disclosed a basalt layer going to the depth of 240 m.

On the flanks of the ridge the thickness of sedimentary cover amounts to 0.2–0.6 km and gradually increases on both sides of the crest. Borehole 114 drilled on the eastern flank passed through 620 m of terrigenous-carbonate deposits dating from Pleistocene to Miocene and reached the basalt basement. In boreholes 407 and 408, drilled on the western flank under a sequence of sediments more than 300 m thick, basal lavas were encountered interlaying with deposits of Middle Eocene and Early Miocene age, which conforms to the length of the distance from the boreholes to the ridge axis [165].

A band of increased sedimentary thickness is traced along the Gibbs Fracture Zone. It is associated with transverse trenches intersecting the rift zone and the ridge flanks and continues on both sides going into the oceanic basins where the thickness of sediments reaches 0.8–1.0 km [203].

On the North-Atlantic Ridge from the Gibbs Fracture Zone to latitude 30° North the observed structure of sedimentary sequence is approximately of the same type. In the rift zone the sediments are practically absent, and on the flanks their thickness is from 0.1 to 0.6 km [137]. Boreholes drilled on the ridge give an idea of the structure and the age of the sedimentary section (Figure 19). Close to the boundary of the rift zone, carbonate silts

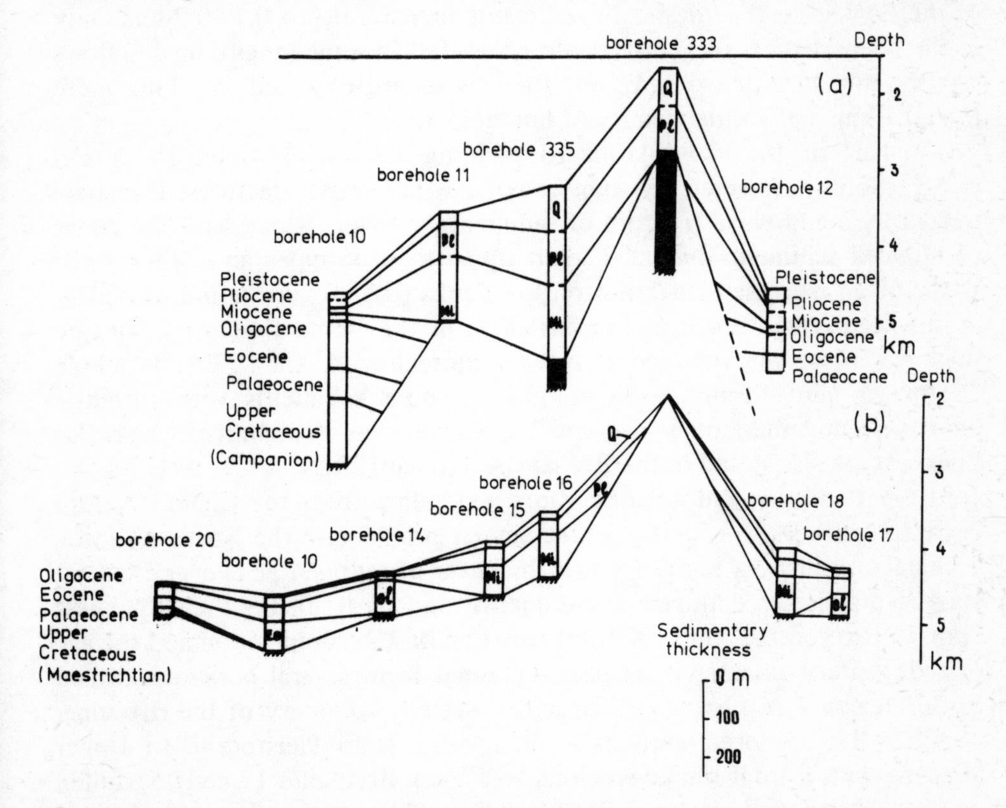

Fig. 19. Sedimentary sequences of the North- (a) and South-Atlantic (b) Ridges according to deep-sea drilling data.

aging from Pleistocene to Miocene have been disclosed, their thickness equal to 280 m (borehole 11). In the vicinity of the outer boundary of the ridge (borehole 10) 456 m of sediments have been drilled through, their upper part being represented by interlayers of red clay and carbonate silts dating from Pleistocene to Eocene, and the lower part – by Upper Cretaceous carbonate deposits with interlayers of volcanic material. In both these boreholes the basalt basement was reached, its upper crust proving to be vitrified. Interesting results were obtained in the rift zone to the south-west of the Azores. Boreholes 332–335, drilled at distances from 34 to 181 km from the rift valley, passed through the sedimentary thickness and penetrated into the basalt basement to depths from 100 to 580 m. The sedimentary thickness increases in the direction away from the rift valley from 104 (borehole 332) to 454 m (borehole 335), and the age of basalt deposits from 3.3 to 11 million years, respectively [165].

To the south, in the zone between latitudes 30 and 15° North, a vast province of reduced thickness of the Mid-Atlantic Ridge sediments is situated. In the rift zone and the adjacent parts of the flanks, sediments are absent or cover the bottoms of valleys with a thin layer. On the outer parts of the flanks the thicknesses of sediments increase up to 0.1–0.2 km. Only in the deepest parts of transverse trenches and in some longitudinal hollows considerably heightened sediment thicknesses are observed, reaching more than 0.5 km, e.g., in the vicinity of borehole 395.

The part of the Mid-Atlantic Ridge lying between latitudes 15° North and 2° South, dissected by numerous large transverse fractures, is characterized by a complex structure of sedimentary cover. Along with the zones of reduced sediment thickness, right up to their disappearance, there exist zones of an increased thickness on the flanks (0.3–0.5 km), and especially in the transverse trenches. For instance, in the Vema Fracture Zone the thickness of sedimentary cover reaches more than 1 km [230]. Borehole 26 passed here through 483 m of terrigenous sediments with turbidite interlayers and did not go beyond the Pleistocene. Sediments of a similar thickness are also noted in the Romanche Trough [23].

The part of the Mid-Atlantic Ridge extending from the Chain Fracture Zone to latitude 30° South, as the similar province in the North Atlantic, is characterized by a relatively low thickness of sedimentary cover. The rift zone is practically deprived of sediments, and on its flanks their thickness does not exceed 0.2 km [137]. Information on the composition and the age of sedimentary thickness has been obtained from several boreholes drilled in this region (see Figure 19). Near the western boundary of the rift zone, borehole 16 disclosed carbonate silts dating from Pleistocene to Upper Miocene with a total thickness equal to 176 m. Boreholes 17 and 15, drilled on the eastern and western flanks of the ridge, passed through 127 and

142 m of carbonate deposits from Pleistocene to Upper Oligocene and Lower Miocene, respectively. The basalt basement was reached in all these cases [165].

The southern part of the South-Atlantic Ridge is less well studied. The structure and the thickness of sediments here are similar to those observed in the northern part of the North-Atlantic Ridge. In the rift zone there are few sediments, and on the flanks their thickness is increasing towards the outer boundaries up to 0.4–0.6 km and more [137]. Apparently, in transverse trenches there are even more sediments.

6. THE RELATIONSHIP OF THE OCEAN FLOOR RELIEF WITH THE BASEMENT RELIEF AND THE ROLE OF SEDIMENTARY COVER IN RELIEF FORMATION

The data obtained by seismic profiling make it possible not only to study the structure and thickness of sedimentary sequences but also to get information on the relief of the underlying basement. These data are of particular importance for investigating the ocean floor and the Mid-Oceanic Ridge where a comparatively small thickness of sediments and a weak influence of denudation factors frequently result in the irregularities of the oceanic basement being directly reflected in the present-day relief. On continental margins, on the contrary, deep submergence of the basement surface, active denudation in subaerial conditions and abrasive-accumulative levelling in the coastal zone and a large thickness of sedimentary cover all mask the relation of the basement structure with the present-day relief and make it indirect, mediated by other factors [57].

As shown above, it is characteristic for the surface of the folded metamorphic basement on the continental margins of the Atlantic Ocean, firstly, that almost everywhere it represents the Mesozoic peneplain. Secondly, owing to the vertical tectonic movements, mainly irregular submergence, the peneplain surface is lowered and curved, forming flexures and projections of the basement diversified by fractures and faults, This surface is overlaid by a sedimentary cover of considerable, if variable, thickness, its layers bedding either monoclinally or subhorizontally. Consequently, the continental shelf is almost everywhere structurally presented by epicontinental platforms morphologically expressed in the form of inclined plains. The processes of abrasive-accumulative levelling had a significant effect on the formation of their relief when the zone of wave abrasion migrated as a result of the irregular submergence of continental margins and the fluctuations of ocean level. This is seen from the presence of discontinuity zones in the sedimentary sequence of continental margins on both sides of the Atlantic Ocean, the most significant of which occurred in the Oligocene and the Miocene.

The structure of the basement surface exerts a more substantial effect on the relief of the continental slope. In the regions of differentiated vertical movements, where the basement on the coast and the shelf is raised, and in the zone of the continental slope lowered along a system of faults, the slope usually has the form of a steep scarp with valley-and-block dissections. The sediments here are accumulated only on steps and at the foot of the slope, so their role in relief formation is not great. In the regions where the basement surface descends along a flexure in the direction of a foredeep and is overlain by a monoclinally bedding sequence of sediments and sedimentary rocks, the continental slope, as a rule, is formed by the outer scarp of this sequence eroded as a result of the action of near-bottom currents, mud flows and slumps. The slope here is represented by a scarp of a concave shape diversified in a number of cases by submarine canyons. An even more considerable levelling of the continental slope is seen in flexures, where a prolonged accumulation of sediments is taking place caused by the active transfer of material from the continent, e.g., in the areas of enormous submarine fans opposite the mouths of the Mississippi, the Amazon and the Niger rivers. The slope here has a gentle surface smoothly passing into the plain of the accumulative series. Prolonged and practically continuous deposition of sediments during the Meso-Cenozoic at the foot of the continental slope has everywhere resulted in the foredeeps, bounding the continental margins of the Atlantic Ocean, being filled up, and it is here that the inversion of submarine relief from the flexures towards the inclined plains of continental rises has occurred.

In the transition zones, where young folded structures have developed, a practically direct conformity is observed between the present-day relief of island arcs and the volcanogenic-metamorphic basement surface. According to geological data [99], island arc ridges are formed owing to the arch-and-block type of uplift of the eroded surface of this basement in the Neogene. Because of the recent age of these structures and the continuing active tectonic processes, sedimentation here plays as yet a negligible role in the transformation of the relief. At the same time, on the bottom of marginal sea basins a quite thick sedimentary cover has accumulated as a result of their prolonged subsidence. Nevertheless, the conformity of the major irregularities in the volcanogenic basement surface to the present-day relief is observed here as well. The sedimentary cover only smoothes out these forms of the primary relief and also conceals the minor irregularities. It is true that in the most submerged parts of the basins the sedimentary cover has reached such thickness that it completely buried the relief of the basement surface, and subhorizontal plains of extreme accumulation have been formed here.

On the Mid-Oceanic Ridge, as shown by the CSPM data, a direct conformity is observed between the oceanic basement and the present-day submarine relief. Sedimentation here is poorly developed and only manifests itself in the filling-up of inter-ridge valleys on the flanks of the ridge and individual depressions in the rift zone. Therefore, the present-day relief practically coincides with that of the oceanic basement. Judging by an increase in the age of rocks and the sedimentary thicknesses on both sides of the ridge axis, the relief has been developing successively from the rift zone to the outer boundaries. This can be explained, based on the plate tectonics concept, by the spreading of the ocean floor away from the axis on both its sides, where the new oceanic crust is being formed practically continuously.

In the zones of abyssal hills the accumulative levelling, although more considerable, is, nevertheless, insufficient to mask completely the primary basement relief. Sedimentary cover here fills inter-ridge valleys, hides the irregularities of the basement surface, and is interrupted on steep slopes of large hills, ridges and seamounts. The present-day relief practically repeats in a smoothed-out form the primary relief of the basement. As one approaches the continental margins, however, the role of accumulative levelling gradually becomes greater. The thickness of sedimentary cover increases, and it successively fills all the depressions, hides the positive forms of the relief and then covers them with subhorizontal layers of deposits. Analysis of CSMP profiles shows that the boundary of the zone of abyssal hills approximately coincides with the sedimentary thickness equal to 500 m, which corresponds, as mentioned above, to the average size of these hills. When the sedimentary cover reaches a thickness of about 1000 m, the hill-and-ridge relief of the basement, becomes completely levelled, so that abyssal plains appear. Only the highest basement rises pierce through the thick sedimentary sequence and come out above the ocean floor in the form of seamounts or volcanic islands.

The situation is different with large positive forms of the relief, such as elevations, plateaus, and ridges. These result from the arch-and-block uplifts of the basement surface, diversified by marginal fractures and faults. In the Mid-Oceanic Ridge zone this surface actually coincides with the present-day submarine relief, and, therefore, the role of sedimentary cover here is insignificant, e.g., on the Azores Plateau or the Palmer Ridge. On the bottom of oceanic basins, especially within the abyssal plains, the basement surface elevations are overlain by a sedimenatry cover of differing thickness, lying on its tops as a kind of 'caps', but on the slopes it either decreases in thickness or is completely discontinued. Thus, the sediments, while smoothing out the primary forms of the relief of large elevations in the oceanic basement, on the whole preserve their integral structure, which is clearly exemplified by the Bermuda and the Rio Grande Plateaus and the Sierra Leone Rise, as well as the Walvis Ridge. Judging by CSPM data, their upper surfaces are mainly covered by undisturbed sediment layers, and this is indicative of calm tectonic conditions, existing there after the beginning of sedimentary cover formation.

CHAPTER 3

The Structure of the Consolidated Crust and Anomalous Geophysical Fields of the Atlantic Ocean

The Earth's consolidated crust serves as the basement for the morphostructures of the ocean floor. Its structure actually determines the tectonics of the Earth's oceanic provinces, and the processes taking place in it exert a substantial effect on the formation of the basic elements of submarine relief. Regional differences in the structure of the consolidated crust and the underlying upper mantle determine the nature and distribution of anomalous geophysical fields – magnetic, gravitational, and thermal, which in this case can be regarded as indicators when studying the structural features of different forms of the ocean floor relief.

1. THE STRUCTURE OF THE EARTH'S CRUST ACCORDING TO SEISMIC DATA

According to seismic data obtained by refraction shooting, the Earth's consolidated crust under the ocean floor consists mainly of two layers: (a) the 'suprabasalt' or the second layer which is the basement for the sedimentary cover (the first layer); (b) the 'basalt', or the third (oceanic), layer. Rocks composing the second layer are characterized by a rather wide spectrum of longitudinal seimsic wave velocities – from 4.0 to 6.0 km s^{-1}, the prevailing velocity values lying within 4.5–5.5 km s^{-1}. Rocks of the third layer are characterized by a narrower range of velocities – from 6.5 to 7.0 km s^{-1} (the average value is 6.7 km s^{-1}), which indicates a greater uniformity of their densities in comparison with that of the rocks of the second layer. The mantle surface at the Mohorovicic discontinuity, underlying the third layer, is usually marked by an abrupt rise in seismic wave velocities, reaching 7.8–8.4 km s^{-1}. The average thickness of oceanic crust is equal to 6–8 km [21, 131].

On continental margins the Earth's consolidated crust has a different structure, typical for continents. Here, under the sedimentary cover lies the so-called 'granite' layer characterized by seismic wave velocities in the range of 5.5–6.4 km s^{-1} (the average value is 6.0 km s^{-1}). Under it lies the 'basalt' layer, similar to the third layer of the oceanic crust in seismic wave velocities, though probably differing from it in composition. The mantle surface here is characterized by the same velocities of seismic waves as those recorded on the ocean floor. On the shelves the average thickness of the continental crust is equal to about 30 km, in sharp contrast to the thickness of the oceanic crust [21].

The above-mentioned common structural features of the Earth's oceanic and continental crust are to a certain extent averaged and typical for the ocean floor and the continental margins taken as a whole. Individual morphostructures have their own regional structural peculiarities that distinguish them from one another. These peculiarities have been brought about by tectonic causes and reflect the history of development of a certain region.

On the continental margins of Precambrian and Palaeozoic platforms, found along the coasts of Greenland, North and South America, Western Europe and Africa, the generally observed structure of consolidated crust is similar to that of the typically continental crust. Local differences are mainly associated with the structure and the thickness of sedimentary cover examined in the preceding chapter. In the continental slope zone the thickness of consolidated crust is everywhere reduced, this reduction mainly affecting the 'granite' layer. In the foredeeps the 'granite' layer is wedging out and nowhere extends into the oceanic crust under the ocean bed. Thus, the boundary of continental (subcontinental) crust extension passes along the foredeeps. From geophysical data, in a number of regions along the inner slopes of these deeps lie zones of vertical (or inclined) boundaries and disturbances, apparently associated with the lines of marginal faults, bounding the continental blocks of the crust (Figure 20).

According to the data of regional geophysical investigations, on the continental margin of Canada and the U.S.A. the consolidated crust consists of two basis layers, their velocity characteristics and thickness changing from one place to another [215]. On the shelf of Labrador velocities in the 'granite' layer are equal to 5.5–5.8 km s^{-1}, in the 'basalt' layer 6.7–7.1 km s^{-1}, but on the continental slope the upper layer is thinning, and under a thick cover of sediments lies a layer with velocity of 7.4 km s^{-1}. In the Newfoundland region the upper part of the consolidated crust section is formed by a layer where the velocities are equal to 6.0–6.7 km s^{-1}, with an intermediate layer where velocities are equal to 7.2–7.5 km s^{-1} found under it. Further to the south-west, along the coasts of Nova Scotia and the U.S.A., the Earth's crust structure is typical for the continental margins. On the shelf the crust, whose total thickness is equal to 33–35 km, consists of sediments and of the 'granite' (velocities 5.7–6.4 km s^{-1}) and 'basalt' (velocities up to 7.0 km s^{-1}) layers. On the continental slope the reduction of the 'granite' layer thickness is accompanied by a decrease in its seismic wave velocities down to 5.3–5.5 km s^{-1}, which is, probably, associated with

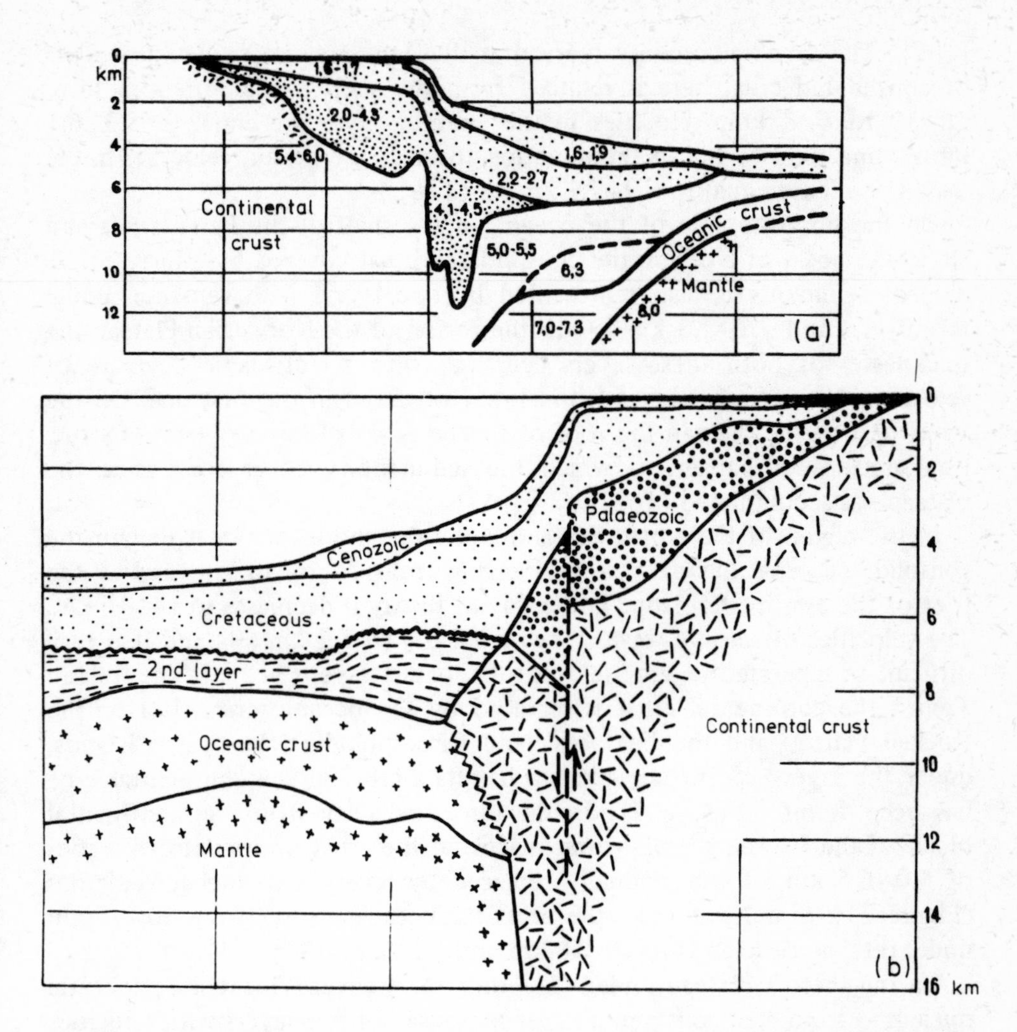

Fig. 20. Seismo-geological sections of the continental margins of : (a) North America; (b) Africa [216]

the decreasing degree of metamorphism of the basement bedrocks with increasing distance from the main Appalachian Orogenic Belt. On the Blake Plateau and the Bahama Platform under thick deposits of limestones two layers with velocities equal to 5.7–6.1 and 7.2–7.4 km s^{-1} have been found. At the foot of the continental slope the first layer is wedging out, and the second is replaced by the third oceanic layer.

Along the coast of Brazil the structure observed is usual for the continental crust: the total thickness of the crust reaches more than 30 km, velocities in the 'granite' layer are equal to 5.8–6.6 km s^{-1}, and in the 'basalt' layer to about 7.0 km s^{-1}. On the continental slope the thickness of the 'granite' layer decreases, velocities in it falling down to 5.3–5.5 km s^{-1}

[185]. The same picture is observed at the Argentina coast. The thickness of conslidated crust here is reduced from the shelf to the foredeep from 28–30 to 6–8 km, velocities in the 'granite' layer are equal to 5.3–6.1 km s^{-1} and tend to fall in the direction towards the ocean, velocities in the 'basalt' layer are equal to 6.9–7.1 km s^{-1} [188].

On the opposite side of the ocean, on the shelf of the Barents Sea and along the coast of Norway the consolidated crust covered by a large thickness of sediments is also represented by two layers with velocities equal to 6.0–6.4 and 7.1–7.3 km s^{-1}. In the region of the Norwegian Plateau the thicknesses of both these layers decrease, with a noticeable decrease of velocities down to 5.8 km s^{-1} and lower observed in the first one. On the outer part of the plateau the roof of this layer rises forming a kind of structural nucleus or barrier, at which the sedimentary cover composing the plateau has accumulated [162].

In the region of the North Sea and along the shores of Great Britain the consolidated crust has a thickness ranging from 24 to 30 km, and at the foot of the continental slope in the Bay of Biscay it decreases to 14–16 km. The velocities of seismic waves range from 5.4 to 7.2 km s^{-1}, and it is very difficult to separate the 'granite' from the 'basalt' layer [112]. In the Ireland Trench the continental crust is replaced by the oceanic crust, but on the Rockall Plateau and the Faeroe-Iceland Rise, including the Faeroe Islands, under the layers of Tertiary plateau basalts a crust of the continental type has been found [113, 207]. In the direction away from the continental block of the Faeroe Islands towards Iceland the crust layers with velocities of 6.0–6.5 km s^{-1} are gradually replaced by layers with higher velocities (Figure 21). Simultaneously, the crust thickness increases from 20–25 km under the Faeroe Islands to 40–50 km under Iceland [27].

On the Iberian Plateau under the sedimentary cover is located a structural nucleus – a block of continental crust composed of two layers with velocities equal to 4.8–5.4 and 6.6–7.1 km s^{-1}. Along the continental slope of the Iberian Peninsula, especially at its foot, a series of faults, bounding the continental block, is traced everywhere [201].

Consolidated crust on the continental margin of Africa has still been investigated insufficiently, but the available data show that here also the crust has a typically continental structure. For example, on the shelf of Sierra Leone the total thickness of the Earth's crust reaches more than 30 km. Besides sediments, the crust consists of two main layers with seismic wave velocities equal to 6.1–6.7 and 7.0–7.3 km s^{-1}. At the foot of the continental slope the thickness of the crust decreases to 12–14 km, the 'granite' layer is cut off by the zone of vertical faults, and the 'basalt' layer, decreasing in thickness, passes into the third layer of the oceanic crust [216].

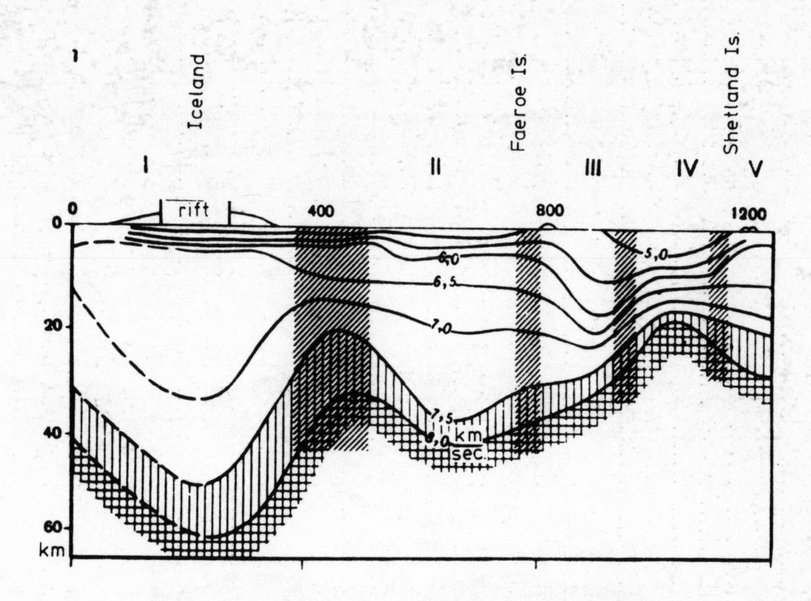

Fig 21. Seismic section of the Earth's crust from Iceland to Shetland Islands [27].

The transition zones are characterized by a complex combination of areas with the continental, subcontinental, oceanic, and suboceanic types of the crust. They differ in both the thickness of the crust and the structure of its basic layers.

In the Mexicano-Caribbean province along the coast of Central and South America extends a narrow belt of the shelf with the thickness of the continental crust equal up to 30–40 km. The same type of the crust is also characteristic for the Nicaragua Rise. Velocities of seismic waves in the upper layer of the consolidated crust are equal to 5.2–5.7 km s^{-1}, in the 'basalt' layer reaching 6.2–6.7 km s^{-1}. For the Antilles island arc and its branches (the Cayman, the Beata, and the Aves Ridges) the subcontinental type of the Earth's crust is characteristic, with a smaller thickness (15–20 km, on the islands of Cuba and Hispaniola up to 30 km) and the presence of a well developed volcanogenic or volcanogenic-sedimentary layer with velocities of 3.0–4.5 km s^{-1}, which is exposed on the islands (Figure 22). The two basic layers are found under it, with velocities equal to 5.0–6.2 and 6.4–6.8 km s^{-1}, respectively [11, 133].

On the floor of the Caribbean Sea basins lies a crust of suboceanic type with a thickness of 15–20 km, and in the Yucatan Basin even less than 10 km. Here, as a rule, three layers are traced under the sedimentary cover; a relatively thin upper layer with velocities of 4.2–5.3 km s^{-1} and two main layers of approximately the same thickness with velocities equal to 5.8–6.3 and 6.7–7.3 km s^{-1}. On the floor of the Gulf of Mexico the crust of

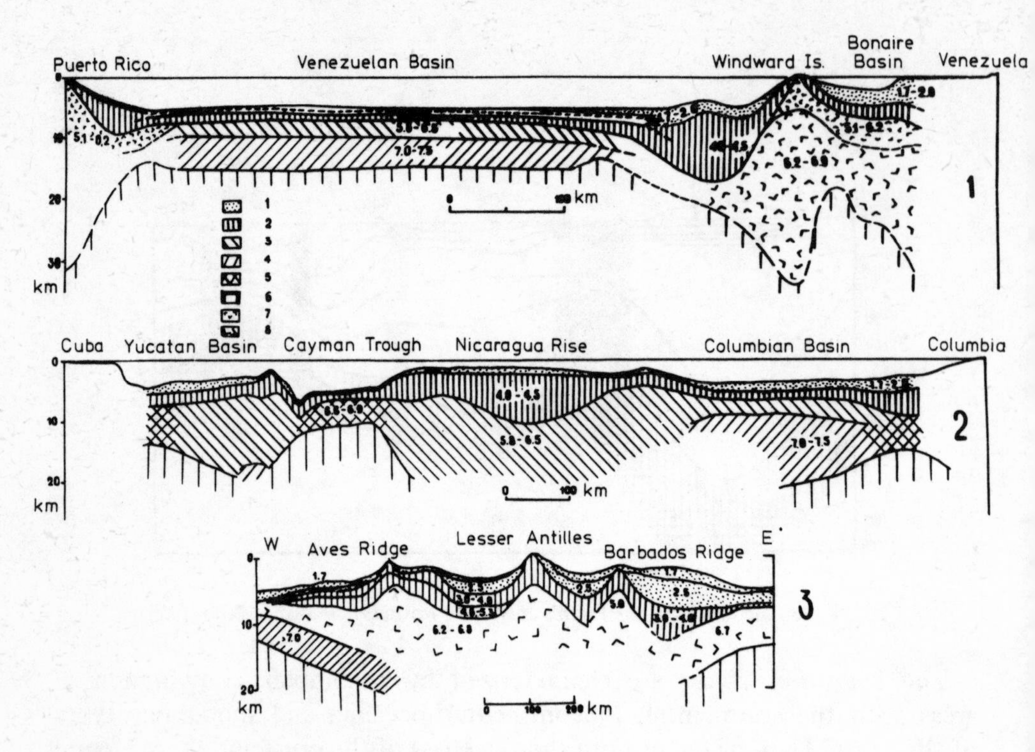

Fig. 22. Seismic sections of the Caribbean Sea [119, 133].

suboceanic type has been also found, only with a greatly increased thickness of sedimentary cover.

Narrow zones of the oceanic crust with a thickness of 6–8 km extend into the deep Puerto Rico and Cayman Trenches. The structure of consolidated crust here is formed by two main layers with seismic wave velocities equal to 5.7–6.3 and 6.6–7.4 km s^{-1}, respectively [118, 133].

The same picture of consolidated crust structure is also observed in the South-Antillean Transition Zone. The northern and southern branches of the Scotia Ridge have subcontinental type of crust of up to 30 km thickness. The crust under the volcanic arc of the South Sandwich Islands and the more ancient arcs located in its rear is somewhat less thick. The upper layer of the crust here is represented by volcanogenic rocks exposed on the islands and characterized by seismic wave velocities of 4.2–5.3 km s^{-1}. Under it are located the main layers of consolidated crust with velocities equal to 6.3–6.6 and 6.9–7.5 km s^{-1}, respectively [96, 134].

The eastern part of the Scotia Sea floor is characterized by a crust of suboceanic type with thickness of about 10–12 km. Here under the sedimentary cover there are layers with velocities equal to 4.2–4.5 and 5.2–6.2 km s^{-1}, underlain by a boundary where velocities are equal to 7.4–7.6

km s^{-1}. The western part of the Scotia Sea and the region of the Drake Passage, on the contrary, are characterized by the oceanic type of the Earth's crust structure. Under a thin sedimentary cover a clearly defined layer with velocities equal to 3.7–4.9 km s^{-1} is traced here, and it becomes markedly thicker below the middle part of the Drake Passage, in the zone of ridges and trenches. A layer with velocities of 6.0–6.8 km s^{-1} is found lower, similar to the third oceanic layer. On both sides of the median zone under this layer a typical mantle boundary has been found with velocity equal to about 8.2 km s^{-1}. The total crust thickness here is equal to 6–8 km [134]. Consequently, the western part of the Scotia Sea should be regarded as the part of the Pacific Ocean floor that has invaded into the South-Antillean Transition Zone, and only the eastern part of the sea floor should be included in the latter.

Oceanic basins, as already mentioned, are characterized by crust of the oceanic type. However, here too there are some local differences between the individual regions. The differences are most significant in the crust of oceanic elevations: ridges, rises, plateaus. All this is indicative of a certain heterogeneity of the consolidated oceanic crust, though on the whole it is more homogeneous than the continental crust.

In the abyssal deeps of the Norwegian-Greenland Basin the consolidated crust of 5–6 km thickness is represented by two layers with seismic wave velocities equal to 4.0–5.4 and 6.8–7.6 km s^{-1}. The seamounts on the floor of the Norwegian Basin are formed by stocks of bedrock with velocities equal to 7.2–7.6 km s^{-1}. In some places the upper layer of consolidated crust has higher velocity characteristics – up to 6.0 km s^{-1} and more [132, 162].

A distinguishing feature of the Laborador Basin, as mentioned above, is the buried Mid-Labrador Ridge. The rocks composing it are characterized by seismic wave velocities ranging from 3.2 to 5.5 km s^{-1}, and their thickness increases from the flanks to the crest from 2.5 to 3.5 km. Under them a layer with velocity of about 7.7 km s^{-1} has been found. In the basin on both sides of the ridge is the usual oceanic crust, where under the sedimentary cover there is a thin second layer (velocities of 4.5–5.3 km s^{-1}) and a considerably thicker third layer (velocities of 6.1–6.9 km s^{-1}). Their total thickness is equal to 6–8 km [73, 125].

In the vast North-Western Atlantic Basin the consolidated crust is almost everywhere represented by two layers with seismic wave velocities equal to 4.0–5.6 and 6.4–7.3 km s^{-1}. The thickness of the second layer varies with the irregularities of the oceanic basement roof. The least thicknesses are found in abyssal plains, where they are equal to 0.5–1.0 km and less. The greatest thicknesses of the second layer, reaching 3–4 km, are recorded on the oceanic rises – the Bermuda Plateau, the Antilles Outer Swell, the

Corner Rise. The underlying third layer also has a variable thickness, but its variations are more gradual. The least thickness is found under oceanic rises, the greatest under abyssal plains. On oceanic rises, in addition to that, another layer with intermediate velocity values of 7.2–7.5 km s^{-1} is sometimes recorded under the third layer [131, 132].

In the Guiana and the Brazilian Basins the velocities of seismic waves in the second layer are equal to 4.8–5.6 km s^{-1}, and its thickness varies within 1–3 km, whereas the third layer of the crust is characterised by velocities of 6.4–7.0 km s^{-1} and the thickness is 3–5 km. On the oceanic Seara and Rio Grande Rises the crust thickness increases because of the increased thickness of the second layer [150].

A similar picture is observed in the Argentine Basin. The consolidated crust here is represented by two layers with seismic wave velocities equal to 4.6–5.0 and 6.4–7.0 km s^{-1}. The thickness of these layers is 0.5–3.5 and 3.0–7.0 km respectively, with the regularities of their variation being the same as in the others basins [188].

In the western and north-western parts of the North-Eastern Atlantic Basin, according to the refraction wave method (RWM) data, the second layer is recorded as absent, though, most likely, it is not very thick and hence is not registered by this method [132]. The thickness of the third layer here varies within 3.5–4.5 km, and the velocities in it are equal to 6.2–6.5 km s^{-1}. As a rule, it is underlaid by a substrate with intermediate velocity values equal to 7.7–7.8 km s^{-1}. In the eastern part of the basin and in the Bay of Biscay, however, there is a marked increase in the second layer thickness up to 2–5 km, velocities in it being equal to 4.2–5.4 km s^{-1}. The thickness of the third layer also increases up to 5 km and more, its velocities reaching 6.6–6.7 km s^{-1}.

In the Canary Basin the observed structure of the consolidated crust is more typical for the ocean floor. The absence or a negligible thickness of the second layer are registered only in some parts of the northern portion of the basin, while on the rest of its territory the thickness of this layer reaches 2–3 km, with velocities in it equal to 6.3–6.8 km s^{-1} [132]. Increased values of crust thickness are recorded under the volcanic massifs of the Horseshoe Rise, Madeira, the Canary and the Cape Verde Islands. Their structure is somewhat unusual and is characterized by the stratification of consolidated crust. For example, on the Canary Islands, under the upper layer of volcanic rocks with velocities of 3.9–4.7 km s^{-1} two other layers have been noted, characterized by velocities of 5.6–6.0 and 7.0–7.1 km s^{-1}. The total crust thickness here is from 12 to 27 km [111].

The Cape Verde, the Sierra Leone and the Guinea Basins, to all appearances, have a similar structure of the consolidated crust, studied in more detail in the Freetown region [216]. The crust here is made up of two layers with seismic wave velocities equal to 4.6–5.4 and 6.4–6.5 km s^{-1}. The thickness of the upper layer varies from 0.5 to 2.5 km, and of the lower one from 2 to 5 km, with the thicknesses noticeably increasing in the direction towards the continental margin and especially under the elevations of the ocean floor. On the Sierra Leone Rise under the second layer has been found a layer with anomalously high velocities (7.0–7.3 km s^{-1}) and a thickness of up to 6 km.

The Angola and the Cape Basins are still poorly investigated. The second layer is known to be well developed here, underlain by a typical third layer [136]. They are both found not only on the basin floor but also on oceanic rises, including the Walvis Ridge where the Earth's crust thickness reaches 22–25 km [92].

The Mid-Atlantic Ridge is characterized by specific structural features of the Earth's crust, distinguishing it from the oceanic basins. Seismic investigations that have been carried out in recent years show that the ridge has approximately the same consolidated crust structure along almost all its length. Only the regions of volcanic massifs, such as the Azores, are somewhat different. The region of Iceland is more sharply distinguished, where the structure of the Mid-Oceanic Ridge is observed to intersect with the subcontinental structure of the ancient 'dry land bridge' between Europe and Greenland.

The upper layer of consolidated crust, whose roof actually forms the block-and-ridge relief of the Mid-Atlantic Ridge, is characterized by seismic wave velocities ranging from 3.4 to 5.8 km s^{-1} (prevailing velocities are equal to 4.5–5.5 km s^{-1}). The thickness of this layer is approximately the same throughout the transverse section of the ridge and is mainly equal to 2–3 km, corresponding to the topographic altitude of the ridge above the oceanic basins (Figure 23). Under the upper layer in the rift zone and the adjoining parts of the ridge flanks is located a layer in which the seismic wave velocities are higher than in the usual third layer, but lower than in the normal mantle – from 7.0 to 7.7 km s^{-1} (prevailing velocities are equal to 7.2–7.5 km s^{-1}). It was impossible to determine its thickness by conventional seismic methods, but by comparison with gravimetric data it has been assumed that in the rift zone the thickness reaches about 20 km. On both sides towards the flanks of the ridge the thickness of this layer seems to decrease. Near the outer boundaries of the ridge it is being replaced by the usual third oceanic layer or subsides under it [131, 220].

However, the described model of the ridge structure, according to new data, requires verification. The results of studying the longitudinal and transverse waves from earthquakes in the rift zone show that the thickness of the layer with velocities equal to 7.2–7.5 km s^{-1} must be of the order of 250 km in a zone 300–500 km wide [114, 200]. This zone forms a

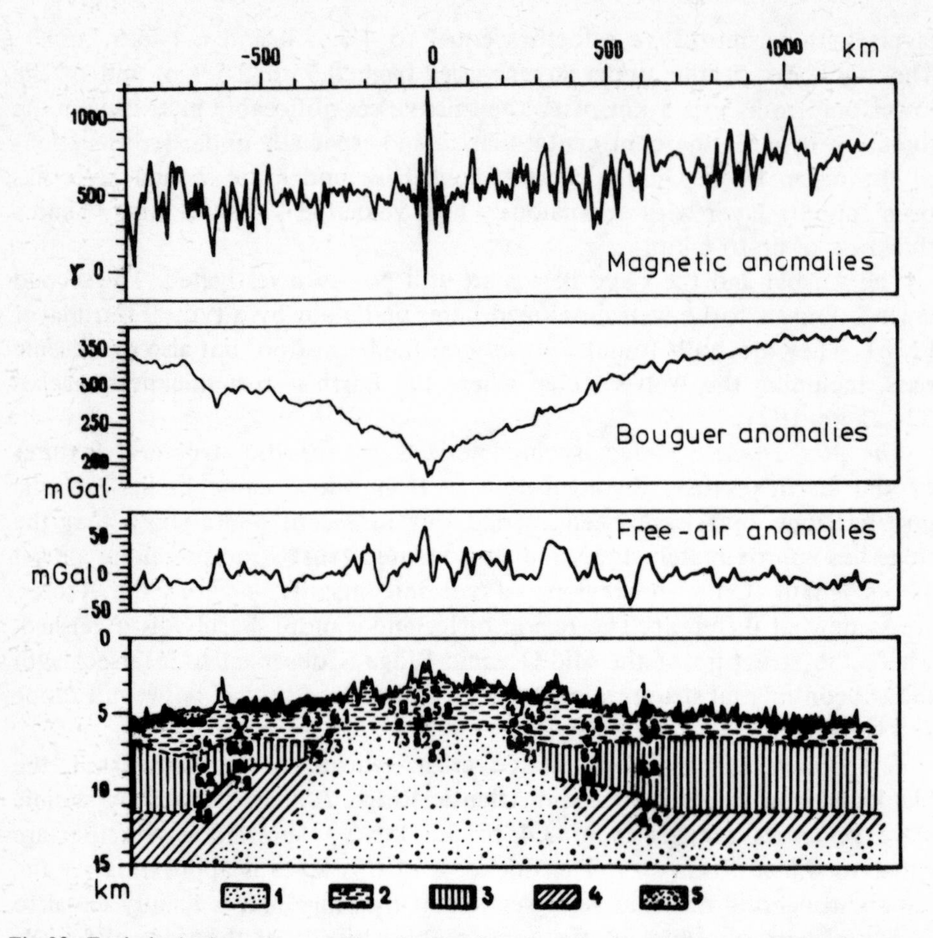

Fig. 23. Typical seismic section, profiles of magnetic and gravitational anomalies of the Mid-Atlantic Ridge [234]. (1) sediments; (2) second layer; (3) third layer; (4) normal mantle; (5) deconsolidated mantle.

barrier for the propagation of transverse waves even from deep-focus earthquakes. On the other hand, the occurrence of shallow-focus earthquakes in the rift zone indicates that the Earth's crust here is sufficiently rigid. From this it follows that the upper layer of the lithosphere in the rift zone of the Mid-Atlantic Ridge of up to 30–35 km thickness (which corresponds to the depth of the foci of shallow-focus earthquakes), including the crust and the substrate (upper mantle), seems to be characterized by the so-called 'dislocational' viscosity, while the underlying layer has 'diffuse' viscosity close to the molten state of matter [94]. The lower boundary of this focal-molten zone under the axial part of the ridge must go down to depths of no less than 200–250 km.

Detailed seismic studies by the deep seismic profiling (DSP) method on polygons, performed by the R/V '*Akademik Kurchatov*', have made it possible to ascertain the structure of the crust in the rift zone and in the rift valley of the ridge [89, 100]. In the regions of the Kurchatov and the Atlantis Fracture Zones, two main layers with velocities of 5.0–5.4 and 7.0–7.2 km s^{-1}, respectively, make up the structure of the crust in the rift zone. The thickness of the first one is about 4 km, but in the transverse trench it decreases to 2.5 km. The thickness of the second layer has been determined only in the transverse trench and the rift valley, where it exceeds 15 km. Under this layer in the rift valley a boundary surface has been found with a seismic wave velocity equal to 8.9–9.0 km s^{-1}, i.e., considerably higher than on the surface of the normal mantle. This, no doubt, is an indication of deep-seated bedrock of greater density rising along axial faults in the rift zone.

Detailed investigation carried out by British scientists at latitude 37° North shows that there is a 2.5 km thick layer extending along the rift zone here, with a recorded velocity of 3.2 km s^{-1}. The relatively low velocity in the layer is explained by the presence of fissures and interstices in it, serving as approach channels for the magma. A layer with velocities equal to 5.4–6.3 km s^{-1} is located below, and at the depth of 36 km the mantle surface has the velocity equal to more than 8 km s^{-1} [236].

In the Azores region as under the other volcanic massifs on the ocean floor, a greater thickness of the Earth's crust is recorded. According to the data from studies of the Rayleigh waves, the crust thickness of the Azores Plateau is greater by 60% than that of the surrounding oceanic crust. The upper mantle here is anomalous and characterized by low velocities of longitudinal waves [213].

In Iceland, where the rift zone outcrops on the land surface, the thickness of the Earth's crust reaches more than 40 km [25]. Under the present-day and the Quaternary volcanic and sedimentary formations there is here a layer of basalts with a thickness equal to 3–5 km, where seismic wave velocities amount to 4.1–6.0 km s^{-1}. A layer with velocities of 6.6–7.0 km s^{-1} lies below, its thickness reaching 30 km. It is underlain by a relatively thin intermediate layer where the velocity is equal to about 7.5 km s^{-1}, under which has been recorded the mantle surface [27]. The anomalously thick crust in the rift zone of Iceland seems to be caused by the fact that here this zone happened to intrude between the blocks of the ancient continental crust of the Greenland-Britain Rise.

In conclusion of our review of the Earth's crust structure from seismic data it should be pointed out that at present, according to the new global tectonics concept, the crust and the upper mantle of the ocean down to depths of about 50–60 km are regarded as a single lithospheric layer (plate). It is very strong in contrast to the underlying asthenosphere layer which is considerably weaker. It is this fact that makes possible the horizontal displacements of lithospheric plates along the asthenosphere surface [84, 184]. The basis of individual plates is comprised of the continents of North and South America, Europe, Africa, and Antarctica (continental plates) to which the adjoining parts of the Atlantic Ocean floor (oceanic plates) are as it were soldered. The existence of such plates is confirmed by seismological data, indicating a marked decrease in seismic waves velocities, especially of transverse waves, at the boundary between the lithosphere and the asthenosphere [166]. The zone of low, in comparison with the normal mantle, seismic wave velocities and low viscosity, which is located under the crest of the Mid-Atlantic Ridge, seems to have been formed as a result of the uplift of asthenospheric diapir and its intrusion into the lithosphere, with the upper part of the layer where the velocities are equal to 7.2–7.5 km s^{-1} representing the deconsolidated upper mantle.

2. GEOLOGICAL NATURE OF THE SEISMIC LAYERS OF CONSOLIDATED CRUST

To understand the geological processes taking place in the Earth's crust and influencing the formation of ocean floor morphostructure it is necessary to know the composition of rocks, making up the basic seismic layers. This composition can be determined both by direct geological investigations of the ocean floor with the help of dredging and deep-sea drilling and indirectly, based on estimating the velocity characteristics of certain rocks and on theoretical concepts regarding the processes of their formation. To determine the composition of rocks making up continental margins it is also possible to extrapolate the data on the geological structure of the adjoining land areas.

On continental margins, as well as on the adjoining land, the 'granite' layer is essentially the basement for the platform sedimentary cover. It is composed of Precambrian and Palaeozoic rocks, crumpled into folds, highly metamorphized and penetrated by granite intrusions. The seismic wave velocities recorded in it are characteristic not only for granites but also for gneisses, crystalline schists, quartzites, marbles and other metamorphic rocks. Geological observations in areas where this folded basement is exposed (on Precambrian shields or in the nuclei of folded mountain structures) confirm the above-mentioned composition of the 'granite' layer rocks. Therefore, from the geological point of view, it should be more correct to designate it as the granite-metamorphic or granite-gneiss layer [7, 98].

The 'basalt' layer is nowhere exposed on the ground surface, so for the present its nature can be estimated on the basis of indirect data only. Judging by velocity characteristics, this layer can be made up of rocks that have undergone the highest degree of metamorphosis in the granulite facies, in contrast to the amphibolite facies of the 'granite' layer metamorphism. Because of this the 'basalt' layer of the continental crust is sometimes referred to as the granulite-basalt layer [7]. On the other hand, the rocks of the lower part of the Earth's crust, which are, evidently, the oldest, must be most of all penetrated by volcanic formations of primary basalt composition. This could determine the considerable role of basalts in forming the third layer of the continental crust [98].

In the transition zones, in the regions of highly developed Cenozoic folding and volcanism, the so-called volcanogenic layer is of great importance. It is developed, as shown above, almost everywhere on the submarine ridges and the floor of the basins of the Caribbean and the Scotia Sea. From the data on the geological structure of islands, where this layer is exposed on the ground surface, and the results of dredging on the submarine slopes of ridges, it would appear that the layer is actually composed of volcanic rocks, its upper part on the ridges and islands being represented by andesites and their varieties and its lower part by the basic rocks of basalt composition [11, 96, 97]. The volcanogenic layer developed on the floor of the basins, as shown by the DSDP data consists exclusively of basalts.

The underlying layer with velocities of 5.0–6.2 km s^{-1}, widespread on submarine ridges and rises of the transition zones and outcropping in the nuclei of anticlinoriums on the islands of the sublatitudinal branches of island arcs, is similar in composition to the granite-metamorphic layer of continental margins and forms here the Upper Paleozoic basement of Cenozoic structures. It is mainly composed of highly metamorphic rocks and the intrusions of igneous rocks penetrating them [97]. The underlying 'basalt' layer of transition zones also seems to be similar to that of continental margins.

On the floor of the basins of transition zones, under the volcanogenic layer, two main layers with velocities of about 6 and 7 km s^{-1} are usually distinguished. They are usually referred to the second and third layers of the oceanic crust [133, 134]. Velocities oberved here are, however, somewhat higher than in the oceanic basins and are close to those recorded in the 'granite' and the 'basalt' layers of the continental crust. On the other hand, the presently available geological data give no indications that metamorphic rocks of continental origin exist on the floor of the basins in the transitory zones.

Most probably, we are dealing here either with rocks of oceanic crust, changed in the course of tectonic development under the influence of the surrounding continental Cenozoic structures or with former continental rocks, changed in the course of the 'basification' of the crust during the subsidence of the basin floor.

On the ocean bed and the Mid-Atlantic Ridge the upper layer of consolidated crust, serving as the basement for sedimentary cover, is made up of volcanogenic rocks, mostly of basalt composition. This is seen from the dredging and deep-water drilling data, as well as the data obtained in investigations on oceanic islands [22, 130, 165, 199, 220]. The prevailing rocks found on the surface of the second layer are tholeiite basalts. Dolerites, diabases, and gabbro are encountered much less frequently. Even more scarce are ultrabasic rocks which have been found only in the zones of transverse faults dissecting the Mid-Atlantic Ridge. Evidently, these rocks are the representatives of the deeper layers of the crust and the upper mantle that have intruded up the faults.

The composition of rocks forming the third layer of the oceanic crust, cannot be explained so definitely, since this layer is practically nowhere exposed on the ocean floor, except for the zones of transverse faults and has not yet been reached by deep-sea drilling, the only exception, perhaps, being borehole 334 drilled in the rift zone to the south-west of the Azores. Here under a 62 m thick basalt layer was found the underlying layer of serpentinized peridotites and gabbroids, penetrated to the depth of 67 m [165]. Taking into account the elastic wave velocities characteristic of the third layer the results of petrographic studies of bedrock specimens brought up from the rift trenches and faults, as well as the theoretical concepts of the process of oceanic crust formation, it can be assumed that this layer is composed of transformed rocks of both the overlying second layer and the underlying upper mantle. Located at a considerable depth in the oceanic crust and subjected to the action of high temperatures and pressures, the rocks of the lower part of the second layer must undergo metmorphosis. As shown by the dredging data, among these rocks are basalts, dolerites and gabbro in the zeolite, green schist and amphibolite facies of metamorphism [198, 234]. On the other hand, in accordance with the hypothesis of plate tectonics, the third layer of the oceanic crust must originate in the rift zones from the rise of deep-seated matter in the form of protrusions and its subsequent expansion in both directions from the axial fault. In the course of this rise the deep-seated rocks, evidently represented by peridotites and other ultrabasic rocks, are subjected to hydration as a result of the interaction with ocean water and converted into serpentinites [85]. Therefore, it would, probably, be more correct to regard the third layer of the oceanic crust as serpentinite-metamorphic.

The composition of rocks in the deconsolidated upper mantle, with seismic wave velocities equal to 7.2–7.5 km s^{-1}, underlying the crust of the Mid-Atlantic Ridge, can be only conjectured from some samples of ultrabasic rocks obtained in the fault zones and from the theoretical concepts of the processes of this ridge development [22, 85]. It seems that here basic magmas are melting out of the primary mantle matter, with lherzolites –

rocks similar to stone meteorites – being taken for its prototype. The ascending basic magmas form the second layer of the crust in the Mid-Atlantic Ridge, and the remainder, after melting out, forms the rocks of peridotite (harzburgite) type. That is why the layer of deconsolidated mantle should, most probably, be regarded as the peridotite layer, and asthenospheric diapir of the focally molten mantle under it – as the lherzolite layer. The described structural model of the crust and the upper mantle in the rift zone is shown in Figure 24.

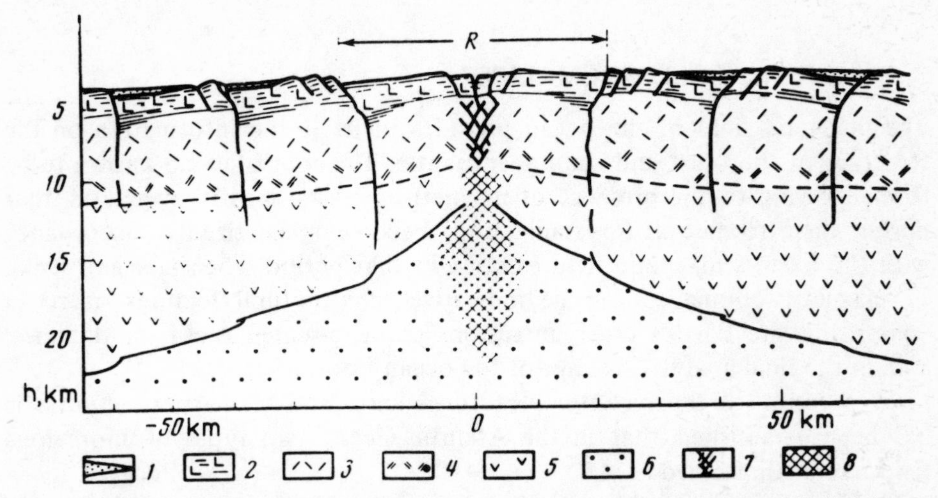

Fig. 24. Structural model of the rift zone [85]. (1) sediments; (2) second layer; (3) third (serpentinite) layer: (4) lower part of serpentinite layer with developed metamorphism; (5) lithosphere; (6) asthenosphere; (7) zone of open fractures; (8) zone of ruptures draining basal liquids from the asthenosphere.

There is also an opinion that the second layer of the oceanic crust is composed not only of basalts but also of sedimentary rocks interbedding with them. One of the arguments adduced is the data of borehole 334, where in the upper basaltic layer small interlayers of consolidated limestone deposits have been found, and in the lower layer of ultrabasic rocks breccias have been disclosed, cemented by consolidated marine deposits of the Miocene age. Proceeding from this, extrapolations are made for the whole second layer or even for the whole consolidated crust of the Mid-Atlantic Ridge, which is assumed to be of volcanogenic-sedimentary origin with the normal bedding of young deposits over more ancient ones [76]. The data cited on the presence of sedimentary interlayers in the upper parts of the section of the Mid-Atlantic Ridge crust can, however, be interpreted in a different way: as the result of local repeated outpourings of basic lavas covering the sediments, or of local tectonic movements involving, together with the volcanogenic basement, a part of the overlying sediments.

On the whole, the geological structure of the Mid-Atlantic Ridge strongly resembles the structure of the ophiolitic suite of continental geosynclinal systems, which may indicate their genetic kinship. Many authors look upon these complexes as the oceanic crust of the geological past [75]. It is quite probable that these present-day continental structures at one time passed through the stage of oceanic development with the formation of mid-oceanic ridges and the spreading of lithospheric plates. These were then involved in a geosynclinal process, subjected to compression and folding, and finally transformed into continental massifs.

3. MAGNETIC FIELD ANOMALIES

The magnetic field of the ocean provides in itself the information on the structure of the crust and the upper mantle. The anomalies are known to be formed owing to the presence of magnetized rocks, mostly magmatic, that during their formation (crystallization) become magnetized in accordance with the Earth's magnetic field existing in that period. The shape and strike of magnetic anomalies can be indicative of structural features of rocks comprising the Earth's crust, intrusions of deep-seated rocks, fault zones, and, in the final analysis, the age of the ocean floor.

As a result of the recently performed aero- and seamagnetic studies it has been established that in the Atlantic Ocean two types of anomalous fields are distinguished:

(1) The linear, or ribbon type, characterized by the predominance of linearly extending alternating positive and negative magnetic anomalies. This type is found on the Mid-Atlantic Ridge and on most of the oceanic basins.

(2) The isometric type, characterized by a combination of mosaically arranged magnetic anomalies with no distinctly expressed strike. It is associated with continental margins and the adjoining parts of oceanic basins.

Though in both types of anomalous fields there exist regional differences in the degree of linearity, wave amplitudes and lengths of individual anomalies, the very occurrence of these types is in principle significant. It reflects profound differences in the nature of magnetic heterogeneity in individual parts of the Earth's crust of the Atlantic Ocean with a differing structure and origin.

Magnetic Anomalies of Continental Margins and Transition Zones

In the region of the Grand Newfoundland Banks several isometric fields of magnetic anomalies with amplitudes of more than 1000 gammas are distinguished. A chain of positive anomalies extends along the shelf indicating the presence of a weakened zone in the Earth's crust into which intrusive bodies have penetrated. The sources of these anomalies are located at great depths, which is in agreement with the great thickness of sediments on the outer shelf. At the same time, intensive magnetic anomalies on the Flemish Cap are caused by the rocks whose upper edge is close to the floor surface [164].

On the shelf of Nova Scotia the magnetic anomalies are stretched along the coast. Close to the coast they have short periods and high amplitudes, and at the edge of the shelf they acquire longer periods and become less intensive. This is, probably, associated with subsidence of the top of the basement made up of magnetized rocks. A chain of magnetic maximums extending along the shelf edge is a part of a marginal magnetic anomaly especially well expressed at the eastern coast of the U.S.A. On the coastal plain and the adjoining part of the U.S.A. shelf the isometric anomalies have the amplitudes of 200–300 gammas, whereas an almost uninterrupted positive anomaly extends along the shelf edge, 30–80 km wide with the amplitudes varying from 1500 to 600 gammas. This marginal anomaly seems to be associated with the uplift of the basement penetrated by intrusions of basic rocks or with the marginal effect in the contact zone between the oceanic and the continental crust of which the coincidence of the maximums of magnetic and gravitational anomalies may be indicative.

In the Caribbean Sea the character of anomalous magnetic field varies from one place to the other [11, 172]. The Yucatan and the Venezuelan Basins are distinguished by a relatively calm magnetic fields, whereas in the Colombian Basin positive and negative anomalies are observed with the amplitudes equal to 150–200 gammas and a wave length from 15 to 40 km. A complex magnetic field with amplitudes of anomalies up to 500 gammas is observed over the Nicaragua Rise. Relatively small magnetic anomalies are traced along the slopes of the Beata and Aves Ridges, and the volcanic arc of the Lesser Antilles. More pronounced positive anomalies with an amplitude of up to 300 gammas are recorded along the arc of the Grand Antilles and along the Windward Islands Ridge to the north of the coast of Venezuela. The greatest anomaly with an intensity of more than 600 gammas is traced along the northern border of the Cayman Trench. At the same time, over the Puerto Rico Trench, magnetic anomalies are considerably less intensive – about 50 gammas – in contrast to the Cayman Trench.

Along the coast of South America, magnetometric investigations have been conducted on the shelves of Guiana, Brazil, and Argentina [138]. Pronounced magnetic anomalies are recorded opposite the Rio de La Plata, near the port of Bahia Blanca and the Golfo San Jorge. The continuation of the Andes crystalline basement on the shelf is traced by magnetic anomalies with an intensity of up to several hundred gammas at Cape Horn.

In the western part of the Scotia Sea the magnetic field is of a linear, banded nature [107]. In the Drake Passage the axes of magnetic anomalies extend in the north-eastern direction, and are symmetrical to the median zone. In the eastern part of the Scotia Sea the axes of magnetic anomalies extend in the sub-meridional direction, parallel to the South Sandwich arc and the submarine swells in its rear.

According to the data of magnetometric surveys, the shelf of Norway and the adjoining marginal plateau are characterized by isometric, low-intensity magnetic anomalies [103]. Only in some places, evidently associated with faults, zones of increased gradients are noted. It is important to stress the substantial difference between the calm magnetic field of the shelf of Norway and the sharply alternating magnetic field of the shelf of South Greenland, caused by the deep submergence of the basement in the former and its closeness to the floor surface in the latter.

A calm magnetic field is also characteristic of the North Sea and the areas along the coasts of Great Britain and France. The magnetic field at the coast of Portugal is of a somewhat more complex form, owing to the presence of marginal fractures and faults. It is of interest to note that the Galicia, the Vigo and the Porto seamounts on the marginal Iberian Plateau are non-magnetic. This is a result of the fact that they are made up of sedimentary or metamorphic rocks [109]. A more complex, sharply alternating magnetic field has been recorded on the Iceland-Faeroes Rise, on the shelf of the Faeroes, and on the Rockall Plateau. This is, evidently, associated with extensive development of the Tertiary plateau basalts having a high but sharply alternating magnetization [207].

Although the continental margin of Africa is less thoroughly studied than that of North America, the data available show a similarity in their magnetic fields [128, 129]. For example, in the region between the Canary and the Cape Verde Islands, along the edge of the shelf lies a clearly defined linear anomaly with the amplitude up to 600 gammas. Its nature is apparently the same as that on the shelf of the U.S.A. A similar marginal anomaly has been recorded further to the south, in the Monrovia region. At the coast of South-West Africa a system of positive magnetic anomalies has also been recorded with amplitudes equal to 200–500 gammas, stretching along the continental slope.

Magnetic Anomalies of the Mid-Atlantic Ridge and Oceanic Basins

As a result of numerous profile surveys and polygon studies it has been established that the Mid-Atlantic Ridge is characterized by a contrasting band-like magnetic field. Its basic features are as follows:

(1) the bands of positive and negative anomalies are orientated along

the ridge strike. Their continuity is only disturbed by transverse tranform faults;

(2) all along the ridge axis is traced the central rift anomaly, its amplitude 1.5–2 times greater than that of the neighbouring anomalies and reaching values of 1000–1500 gammas in the temperate and polar latitudes;

(3) the rift anomaly is the axis relative to which the other anomalies are symmetrically located on both its sides;

(4) a regular alternation of anomalies with a characteristic structure is recorded.

This makes it possible to correlate individual anomalies even with a widely spaced network of profiles.

On the basis of the above principle, characteristic symmetrical anomalies have been distinguished with the numbers from 1 to 32, beginning from the rift anomaly in both directions away from the ridge axis, assigned to them [158]. Apart from the basic anomalies, additional ones, less marked and not conforming to the general scheme, have been identified. They are, apparently, indicative of local dislocations in the oceanic crust.

The Reykjanes Ridge is one of the areas where the banded structure of magnetic field is clearly manifested. The axial and lateral anomalies, extending along the whole ridge and coming out into the shelf of Iceland, are clearly seen here. The widths of positive and negative anomalies on both sides of the axis are proportional to the duration of the epochs of palaeomagnetic polarity, known to have repeatedly changed during the Meso-Cenozoic. Wide bands of anomalies correspond to the epochs of normal (Brunhes, Gauss) and reversed (Matuyama, Gilbert) polarity, and narrow bands of anomalies correlate with the paleomagnetic intervals of these epochs: Jaramillo, Olduvai, Kaena, Mammoth *et al.* [221].

A banded magnetic field has also been found in the Labrador Basin, associated with the Mid-Labrador Ridge buried under the sediments. In the south-west this system of banded anomalies connects with the Mid-Altantic Ridge anomalies [178].

In the region of Iceland and to the north of it the picture of the anomalous magnetic field is rather complicated. According to the data of investigations of R/V *Akademik Kurchatov* and to foreign sources [103, 196] the Iceland, Mohns and Knipovich Rift Ridges are characterized by clearly defined banded anomalies, orientated along their strikes. In the Norwegian Basin the strike of anomalies is seen to be sub-meridional, coinciding with the orientation of the seamounts chain. On the other hand, the Jan Mayen Ridge and the Iceland Plateau are characterized by mosaic, sign-variable fields with the intensity equal to 200–400 gammas, which may indicate the subcontinental type of the crust structure. In the zone of the Central Graben of the island, and partially in its eastern and western areas, there is a series of banded anomalies comparable to the banded anomalies of submarine rift ridges.

The North-Atlantic and the South-Atlantic Ridges are characterized almost throughout by a rather clearly defined banded magnetic field [80, 89, 124, 204]. The only exception is the region of Equatorial Atlantic, where, near the magnetic equator, because of the peculiarities of the Earth's magnetic field, the anomalies are very poorly expressed, and it is, therefore, impossible to identify and trace them over a long distance. Having summarized the work of numerous expeditions American authors have compiled a summary map of the magnetic anomalies of the ocean [205], supplemented by the data obtained in the studies on board R/V *Akademik Kurchatov* and by other data (Figure 25). Comparing the geomagnetic profiles and the map of magnetic anomalies with the ocean floor relief, one can see that anomaly 5 corresponds approximately to the rift zone boundaries, the outer boundaries of the Mid-Atlantic Ridge in the northern part of the ocean (to the north of the Gibbs Fracture Zone) coincide with anomaly 20, and in the central and southern parts of the ocean with anomalies 25 and 26.

Comprehensive studies on polygons make it possible to ascertain the morphology and the structure of banded anomalies and their interrelation with the ocean floor morphostructure. For example, on the polygons in the regions of latitudes 45° and 40–41° North the rift zone is represented by an arch-like uplift with a deep rift valley and relatively shallow dissected rift ridges stretching over considerable distances. The magnetic anomalies here have a well-defined linearity and large gradients – about 200 gammas km^{-1}. But on the polygons located further to the South, in latitudes 30°, 27°, 22–23° North and 6–8° South, the rift zone relief is dissected into large blocks, and the ridges are not so extensive. The magnetic anomalies are correspondingly less linear and less contrasting, with gradients of about 100 gammas km^{-1}. The magnetic field in these regions resembles a mosaic one, but, at the same time, individual anomalies are situtated in chains one after the other, forming, on the whole, intermittent bands [64, 187, 228, 229].

Using the method of transforming the field into the lower half-space it has been established that banded magnetic anomalies observed in the rift zone of the ocean floor surface represent the overall effect of the local, short-period, highly intensive anomalies, associated with the Earth's crust irregularities whose horizontal dimensions are equal to 0.5–3.0 km. The cross-sectional dimensions of these local anomalies turn out to be close to those of the relief forms of the second order diversifying the basic structures of the rift zone: valleys, blocks, and ridges. As shown by calculations, the upper edges of magnetically disturbing bodies are located here in the immediate vicinity of the floor surface [89].

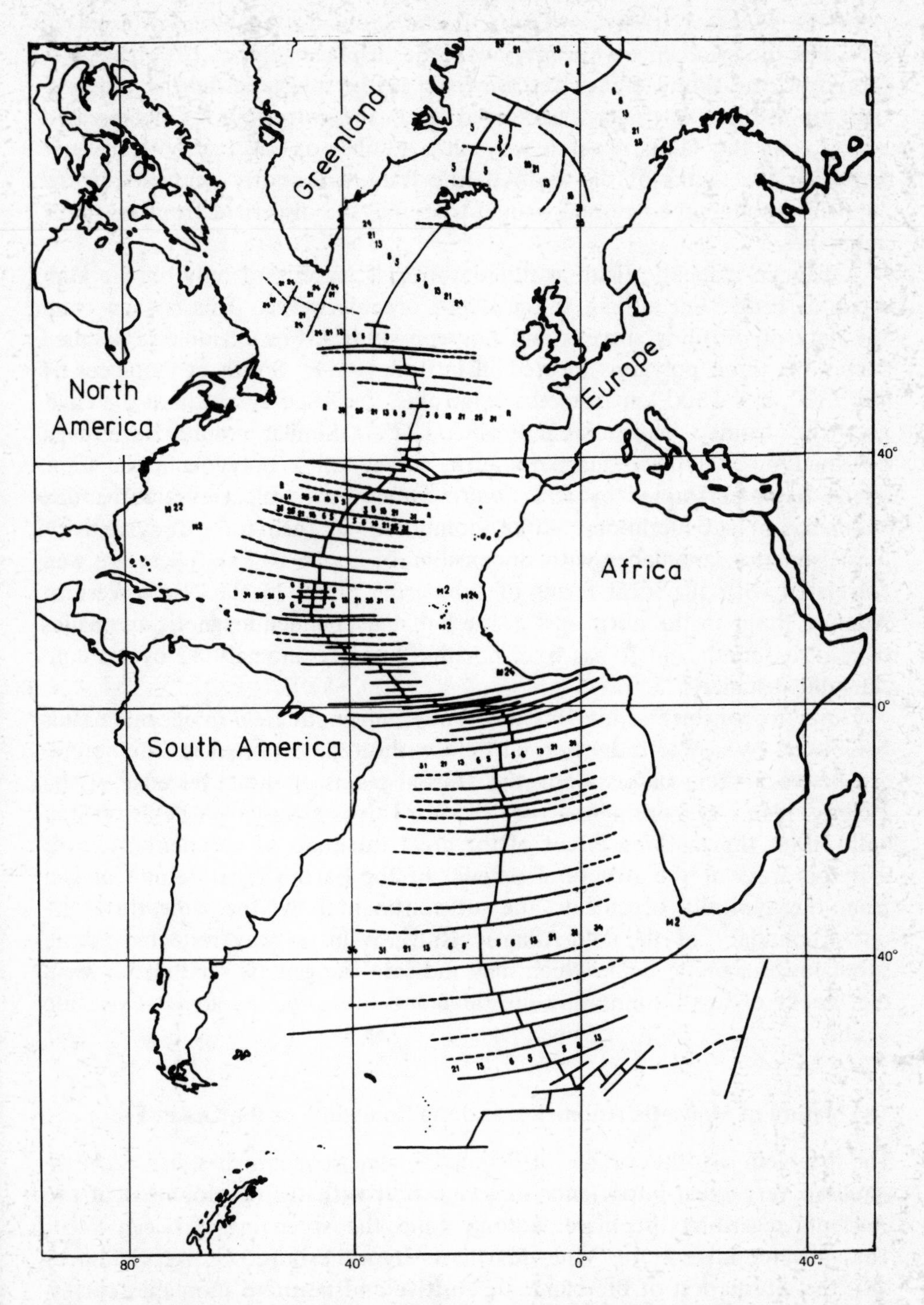

Fig. 25. General scheme of magnetic anomalies in the Atlantic Ocean [204, 205], supplemented.

In the zones of transverse faults, linear anomalies are usually displaced in either direction in conformity with the displacement of rift structures. Moreover, the faults themselves are fixed in the magnetic field as independent anomalies, obviously associated with the intrusions of deep-seated rocks along the faults. That is why, in a number of cases, when transverse faults on the flanks of the Mid-Atlantic Ridge are poorly expressed in the submarine relief, their presence and strike may be discerned from magnetic anomalies.

A banded magnetic field, as noted above, is traced not only on the Mid-Atlantic Ridge but also in the adjacent oceanic basins. This is seen from the data of profiling surveys and polygon studies. For example, a detailed survey on three polygons located in latitudes 6–8° South at distances of 200, 800, and 1300 km from the ridge axis, disclosed everywhere the characteristic banded magnetic anomalies [229]. Similar results have been obtained by a detailed magnetometric survey on a polygon in the Cape Verde Basin performed by R/V *Dmitri Mendeleev*, which revealed banded anomalies of sub-meridional strike, composed, in their turn, of systems of local isometric anomalies with dimensions of about 10 km. These are well correlated with the local forms of submarine relief. In the North-Western Atlantic Basin to the north-east of the Bahamas, banded magnetic anomalies were also found, and it has been possible to trace anomaly 32 over a considerable distance [235].

Closer to continental margins the banded magnetic field in oceanic basins is replaced by a practically normal field without any anomalies. The boundary between them passes along the abyssal plains or the outer edge of the inclined plains of continental rises [157]. This is assumed to be associated with either the masking effect of the great thickness of sedimentary cover in these areas or the structural scheme of the Earth's crust being changed from the typically oceanic to the sub-continental. At the same time, the outer boundary of the Mid-Atlantic Ridge is in no way reflected in the anomalous magnetic field, which may indicate the genetic relation between the blocks of crust composing the ridge and the adjoining areas of oceanic basins.

The Nature of Magnetic Anomalies and the Spreading of the Ocean Floor

The problem of the origin of banded magnetic anomalies has recently acquired very great importance in connection with the development of the concepts regarding the plate tectonics and the spreading of ocean floor. The most popular is the Vine–Matthews hypothesis [232]. According to this, the alternation of the bands of positive and negative anomalies corresponds to the polarity reversal in the magnetically active layer of the Earth's crust, formed as a result of the molten deep-seated matter entering into the axial zone of the Mid-Oceanic Ridge, its cooling, crystallization, and spreading in both directions from the axis under the conditions of periodical reversal of the geomagnetic field. This leads to the formation of the alternating blocks of oppositely magnetized bodies responsible for the anomalies in the oceanic crust.

The above hypothesis explains quite satisfactorily all the characteristic features of banded anomalies: their homogeneity, extent, regular alternation, and symmetry relative to the ridge axis. Proceeding from this hypothesis, it is to be assumed that the history of geological development of the ocean floor is fixed in magnetic anomalies [36]. If the hypothesis is valid, it becomes possible to reproduce the geological history of the ocean floor from the magnetometric data.

The existence of geomagnetic field reversals, inferred from the data of studying the magnetization of lava flows on land, is regarded as proved, at least for the last 4 million years [123]. Based on this, by extrapolating the reversal scale to all the banded magnetic anomalies found in the ocean, the geomagnetic polarity time scale has been developed to cover the period of about 160 million years. The validity of this scale has been quite satisfactorily confirmed by the data of deep-sea drilling. According to them, the age of rocks on the surface of oceanic basement usually coincides with or is rather close to the age of magnetic anomalies located in this region.

The measurements of residual magnetization of igneous rocks in the rift zone of the Mid-Atlantic Ridge, performed by R/V *Akademik Kurchatov* [77], have shown that positive anomalies can be associated both with fresh and metamorphized basalts, with serpentinized peridotites, and the magnetic varieties of gabbro. Negative anomalies can only be created by unchanged basalts having the highest values of residual magnetization. Residual magnetization of basalts has been found to decrease in the direction away from the rift zone which is, probably, caused by the oxidation of titanomagnetites, the degree of this oxidation depending on numerous factors, among them the rate of the ocean floor spreading, i.e., the remoteness of basalts from the volcanic source. This decrease in the residual magnetization of metamorphized basalts allows one to maintain that basalts on the floor surface serve as the sources of anomalies only in the rift zone. On the flanks of the Mid-Oceanic Ridge, only the slightly metamorphized basalts or the deep-lying ultrabasic rocks can serve as the sources of anomalies [69].

At the same time, calculations of the magnetic effect exerted by the upper layer of basalts, responsible for the anomalies, show that the calculated and the observed curves do not always coincide, which is indicative of a considerable magnetic heterogeneity of the Earth's crust. Facts have been recorded of reversed magnetization of basalts in the axial part of the rift zone, not coinciding with the present-day polarity, and of a stratified structure of basalt masses where the different-age layers date from different palaeomagnetic epochs. The boundaries between the basalt layers of different epochs are situated at depths of 2.5–3.0 km below the floor surface. All this casts doubt on the absolute correctness of the Vine-Matthews hypothesis and the soundness of the concepts on the chronology of banded magnetic anomalies, though it does not disprove the concept of the ocean floor spreading as a whole [77].

Based on the analysis of magnetic anomalies, the rate of the Atlantic Ocean Floor spreading has been estimated. In the South Atlantic, e.g., the rate of spreading in the rift zone proved to be equal to about 2 cm yr^{-1}, at latitudes 22–23° North 1.4 cm yr^{-1}, at 30° North 1.9 cm yr^{-1}, at latitudes 40 and 45° North 1.2 cm yr^{-1}, on the Reykjanes Ridge about 1.0 cm yr^{-1} [64, 124, 221]. If one analyses magnetic anomalies at the flanks of the ridge, dating from earlier stages of geological development, the rates of ocean floor spreading will turn out to be different: either higher or lower than the present-day ones. This shows that with the passage of time the spreading process went in an irregular fashion, at times accelerating, at other times slowing down, which is generally true for many tectonic processes taking place in the Earth's crust.

In terms of the cyclic development of spreading ridges, one can explain the heterogeneity of rift zone structure and the differences in the structure of magnetic field described above. The northern regions of the Atlantic Ocean obviously correspond to the initial stage of the last cycle, and the southern ones to more recent stages when intensive vertical movements are taking place. The intrusion of a large number of dykes and the effusions of tholeiite basalts at the initial stages can be assumed to have resulted in the formation of elongated crests widespread in the northern regions. Their association with the local magnetic anomalies indicates that they are composed of volcanogenic rocks with a high magnetization. At the later stages of the cycle the formation of faults leads to the development of large block structures characteristic for the southern regions. This also causes modifications in the structure of magnetic anomalies, less linear and not so contrasting in the southern part of the ocean.

4. GRAVITY ANOMALIES

Gravity anomalies contain information on the density nonuniformities in the Earth's crust and the upper mantle. Usually, when interpreting geologically the data of gravimetric surveys, the anomalies are examined in reduced form as free-air, Bouguer, and isostatic anomalies. Each of these reductions has its physical sense, its merits and demerits. The first two have found the most extensive application at present [13, 21]. The free-air anomaly is obtained directly in a gravimetric survey on board a vessal located at the ocean level, the Bouguer anomaly is calculated with a correction for the ocean depth and the floor relief.

The averaging of free-air anomalies on 5° X 5° squares performed by American investigators [219] has revealed considerable positive anomalies over the Mid-Atlantic Ridge, the Walvis Ridge and other large ocean floor elevations, their amplitudes being equal to +20 mGal and extending up to 1500 km, and negative anomalies over the oceanic basins, with amplitudes of more than 30 mGal and extending more than 1500 km. In a number of regions, zones of positive anomalies with amplitudes reaching +80 mGal and more have been identified. A vast area of considerably large positive free-air anomalies is the region around Iceland and to the north of it. Large positive anomalies have been recorded in the Azores region.

Continental margins and transition zones, oceanic basins, submarine ridges and elevations are also clearly distinguished by the calculated Bouguer gravity anomalies [14].

Continental Margins and Transition Zones

Continental margins are characterized by certain common features, most comprehensively studied at the eastern coast of the U.S.A. [127]. The Bouguer anomalies, when passing from the continent to the ocean, as a result of the reduction of crustal thickness, increase from –50–0 mGal along the coast up to +300 ÷ 325 mGal on the ocean floor, forming a narrow belt of large horizontal gradients. The free-air anomalies form a characteristic belt of conjugated positive (+60 mGal) and negative (–60 mGal) anomalies caused by the mutual influence of the marginal effect and a thick sedimentary sequence on the shelf and the continental slope.

A similar picture is observed at the continental margins of Africa and Western Europe, where the averaged Bouguer anomalies increase towards the ocean from 0 to +280 mGal in a belt 250–300 km wide. However, in the regions of the Norwegian Marginal Plateau, the Iceland–Faeroes Rise, the Rockall Plateau and, evidently, other marginal plateaus, the areas with the Bouguer anomalies expand, and the anomalies here are from +80 to +120 mGal. A belt of negative free-air anomalies is traced along the continental slope of Scandinavia, along the Faeroes – Shetland and the Iceland Trenches and continues further to the south, indicating the presence of a foredeep [12].

In the South Atlantic the continental margins are also characterized by large gradients of gravity force anomalies in the Bouguer reduction and the appearance of a narrow belt of negative free-air anomalies over the continental slope. The latter is absent only on the part of the continental slope in the region of the northern coast of Brazil, where, as one passes from the ocean to the shelf, anomalies increase up to + 80 mGal, and then decrease towards the shore to + 5 mGal. The Bouguer anomalies here are positive – up to 100 mGal [14].

A very complex anomalous gravitational field is observed in the Caribbean and the South-Antillean Transition Zones. This is caused by a combination of formations with a differing crustal structure and a considerable dissection of submarine relief [13, 172]. The basins are characterized by small positive or negative free-air anomalies and very high Bouguer positive anomalies. For example, the free-air anomaly in the Venezuelan Basin is equal to about – 20 mGal, in the Colombian Basin from – 4 to + 26 mGal, in the Yucatan Basin to about – 6 mGal, in the eastern part of the Scotia Sea from + 20 to + 40 mGal. Only in the Drake Passage is a marked increase in the free-air anomaly, up to + 180 mGal, recorded. The Bouguer anomalies in the Caribbean Sea basins range from + 160 to + 280 mGal, in the eastern part of the Scotia Sea from + 250 to + 300 mGal, and in the Drake Passage they rise to + 360 mGal. The marginal parts of the basins are also characterized by narrow zones of negative free-air anomalies, extending along the feet of the slopes in the sublatitudinal branches of island arcs, which indicates the presence of longitudinal marginal fractures.

Island arcs, ridges and rises in the transition zones are characterized by moderate positive free-air anomalies (from + 50 to + 150 mGal) and positive Bouguer anomalies (from + 50 to + 200 mGal). Here the gravitational anomalies, however, undergo rather sharp local fluctuations associated with the complex structure of these elevations. For example, the free-air anomalies increase noticeably over islands, especially over volcanic islands – the Lesser Antilles and the South Sandwich Islands (to more than + 200 mGal), and decrease in the straits between them.

Oceanic trenches are characterized by considerable positive Bouguer anomalies (up to + 330 mGal) and large negative free-air anomalies. Incidentally, it is with the Puerto Rico Trench, that the highest negative free-air anomaly for the Earth's surface is associated: up to – 380 mGal [114]. The axis of the zone of negative anomalies does not coincide with the axis of the Puerto Rico Trench, but is somewhat displaced towards its southern slope, which seems to be explained by the inclined position of the plane of a deep-seated fault (the Benioff fault plane) along which the trench lies. Rounding the arc of the Lesser Antilles, a zone of negative anomalies passes in the south of the Puerto Rico Trench into the Barbados Ridge, thus stressing the genetic relationship between them (Figure 26).

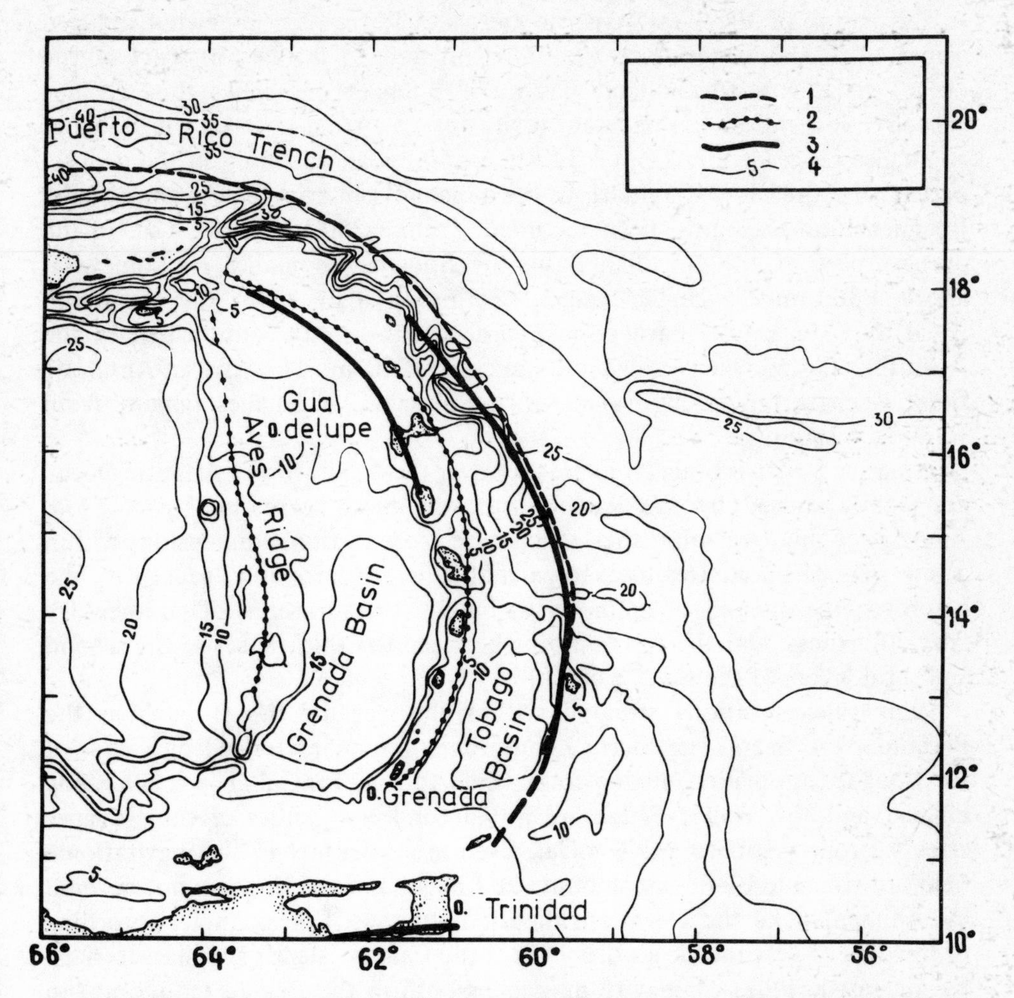

Fig. 26. Scheme of the strike of gravitational anomalies in the Lesser Antilles arc region [119]. (1) axes of negative free-air anomalies; (2) axes of positive free-air anomalies; (3) axes of the minima of Bouguer anomalies; (4) isobaths in hundreds of fathoms.

Oceanic Basins

Oceanic basins with depths equal to 3000–6000 m are characterized by large positive Bouguer anomalies [14]. The deepest part of the Canary Basin is distinguished by enhanced values of the Bouguer anomalies, equal up to +420 mGal. The Iberian and the North-Eastern Atlantic Basins have values of the Bouguer anomalies equal up to +340 + 370 mGal. In the Cape Verde, the Sierra Leone, the Guinea, and the Angola Basins the maximum values of the Bouguer anomalies are equal to +386, +338, +348, and +358 mGal, respectively.

The basins of the West Atlantic are characterized by somewhat reduced values of the Bouguer anomalies as compared to the eastern part of the ocean. In the Newfoundland Basin the Bouguer anomalies thus do not exceed +356 mGal. In the vast North-Western Atlantic Basin the Bouguer anomalies range mainly from +320 to +360 mGal, and only in the deepest part do they reach +426 mGal. In the Guiana Basin even lower values of the Bouguer anomalies have been recorded: from +280 to +320 mGal. In the deepest part of the Brazilian Basin the Bouguer anomalies are somewhat above +360 mGal. Higher values of this anomaly (with the maximum equal to +396 mGal) have been recorded in the south-western part of the Argentine Basin, where the depths exceed 5500 m. The African-Antarctic Basin is characterized by values of the Bouguer anomalies ranging from +320 to +360 mGal.

Intensive positive Bouguer anomalies over the basins of the Atlantic Ocean are mainly caused by the Earth's crust thickness decreasing to 6–7 km. Analysis of the local anomalies which have cross-sectional dimensions within a few tens of kilometres has shown that they are, probably, caused by the heterogeneity of density in the crust blocks, the variations of sedimentary cover thickness, and the relationships between the thicknesses of the second and third layers of the Earth's crust [14].

Submarine elevations situated within the oceanic basins, such as the Bermuda, the Seara, the Sierra Leone Rises, are characterized by values of the Bouguer anomalies ranging from +260 to +300 mGal. The Rio Grande Plateau and the Walvis Ridge are noted for lower values of the Bouguer effects: from +160 to +240 mGal. Even more distinct in the gravitational field are volcanic islands, seamounts and massifs, since they are not covered by sediments like the elevations mentioned above. For example, according to the data of gravimetric studies [111], the Canary Islands are characterized by increased values of gravity anomalies both in the free-air reduction (up to +200 mGal as against −50 mGal over the adjoining part of the ocean) and in the Bouguer reduction (+250 mGal over the islands and +130 mGal over the ocean floor).

The Mid-Atlantic Ridge

All along its length (from the north to the south) the Mid-Atlantic Ridge is distinguished in the gravity field by reduced values of the Bouguer anomalies, as compared to the adjoining oceanic basins [14]. Thus, over the Reykjanes Ridge the Bouguer anomaly ranges from +120 to +160 mGal, over the North-Atlantic Ridge from +200 to +240 mGal, over the South-Atlantic Ridge from +160 to +200 mGal, and over the Atlantic-Antarctic Ridge it is about +140 mGal. Characteristic is the superposition on the Mid-Atlantic

Ridge gravitational field of the sublatitudinal anomalous zones, diversifying this field. In the region of the Greenland-Britain Rise the anomalous field of the ridge with values of up to + 160 mGal is superimposed by the zone of reduced values of the Bouguer anomalies of from 0 to – 30 mGal over Iceland and from +40 to +80 mGal over the Greenland-Iceland and the Iceland-Faeroes Rises. In the region of the Azores the Bouguer anomalies go down to +110 mGal. In both directions from the Azores Massif the zones of reduced values of the Bouguer anomalies extend beyond the ridge, reaching the Grand Newfoundland Banks in the west and the Strait of Gibraltar in the east. Other transverse zones of reduced values of the Bouguer anomalies are outlined in the Equatorial Atlantic (between the Seara and the Sierra Leone Rises) and the South Atlantic (between the Rio Grande Plateau and the Walvis Ridge).

Polygon investigations carried out in recent years on the Mid-Atlantic Ridge make it possible to describe in greater detail the gravity field of the rift zone and to note the differences between individual regions [89]. For example, in the region of latitude 41° North, the average value of the free-air anomaly reaches + 50 mGal, while at latitude 30° North it does not exceed +30 mGal. There are also differences in the Bouguer anomalies, apparently caused by the heterogeneity of the Earth's crust structure.

The density of the samples of basic and ultrabasic rocks, obtained by dredging, varies from 2.5 to 2.9 g cm^{-3}, depending on the degree of serpentinization and dynamic metamorphism. On the whole, one can note a certain correlation between enhanced and reduced values of the Bouguer anomalies and outcrops of rocks with an increased and a reduced density on the ocean floor. The local Bouguer anomalies seem to be largely caused by the influence of crust blocks of different density, which is confirmed by quantitative estimations [89].

By correlating the gravity anomalies, observed over the Mid-Atlantic Ridge and the seismic data on the presence of rocks with the velocities of longitudinal waves (P waves) equal to 7.2–7.5 km s^{-1} on top of the second layer, it was previously assumed, as already mentioned, that the body formed by these rocks has a lens-like shape with a thickness of up to 30 km [220]. The presently accepted concept is that the above body is the roof of the deconsolidated upper mantle, formed above the dome of the rising asthenospheric diapir. Special calculations show that deconsolidation of rocks in the upper mantle by 0.05 g cm^{-3} under Iceland is recorded down to the depths of 120–150 km [112]. The Mid-Atlantic Ridge in the region of latitude 46° North, according to seismic and gravimetric data, is also assumed to lie on an anomalous mantle with densities lowered by 0.04 g cm^{-3} down to the depth of about 200 km [173].

5. HEAT FLOW THROUGH THE OCEAN FLOOR

As a result of different physico-chemical processes in the interior of the Earth, thermal energy is constantly released, and brought to the surface in the form of heat flows. Numerous measurements made on the continents and in the oceans have made it possible to establish that the average heat flow for the whole of the Earth is equal to about 1.5 μcal cm^2 s^{-1}, and if one excludes the zones with anomalously high heat flow the average value will decrease to 1.2 μcal cm^2 s^{-1} [82]. Heat flow distribution is, however, irregular and depends on the geostructural features and the tectonic development of a certain region.

Direct measurements of heat flow through the Atlantic Ocean floor have by now been made in many regions, but not everywhere. The best studied are the Mexican-Caribbean Transition Zone, the floor of oceanic basins, and the Mid-Atlantic Ridge between latitudes 50° North and 40° South. A series of measurements has been made during the cruises of R/V *Akademik Kurchatov*.

The Gulf of Mexico with its extremely thick sedimentary cover is characterized by an anomalously low heat flow of 0.83 μcal cm^2 s^{-1}. However, in the zone of the Sigsbee Knolls, expressed in the relief in the form of salt domes, the heat flow is higher up to 2.1 μcal cm^2 s^{-1}. On the Caribbean Sea floor the heat flow is close to the global average – from 1.1 to 1.4 μcal cm^2 s^{-1}. A somewhat higher value (1.4–1.5 μcal cm^2 s^{-1}) has been recorded on the submarine Cayman, Beata, and Aves Ridges, which is, most probably, caused by less thick sedimentary cover, playing here the role of a screen since the heat conduction of sediments is relatively low. Heightened values of heat flow (up to 2.0 μcal cm^2 s^{-1}) are recorded all along the Lesser Antilles island arc, which is associated with volcanic activity. The same heat flow is observed in the Cayman Trench, while the Puerto Rico Trench is characterized by low heat flow values (1.1–1.2 μcal cm^2 s^{-1}) [11].

The average heat flow value for oceanic basins is equal to about 1.2 μcal cm^2 s^{-1}. Detailed examination of the measurement data shows, however, that the marginal parts of the basins with a heightened thickness of sedimentary cover, as well as the continental rise regions, are characterized by lower heat flow values ranging from 0.6 to 1.2 μcal cm^2 s^{-1}. In the zones of abyssal hills a certain rise in heat flow values is noted – from 1.3 to 1.4 μcal cm^2 s^{-1} – though low values are also encountered here [78]. It should be noted that oceanic elevations, such as the Bermuda Plateau, the Seara and the Sierra-Leone Rises, and the Rio Grande Plateau according to the data available, are characterized by the same heat flow values as those on the floor of the basins.

The Mid-Atlantic Ridge region differs from the oceanic basins, firstly, in a large dispersion of heat flow values and, secondly, in anomalously high values in the rift valley and the transverse trenches. The former seems to be explained by great complexity of the ridge structure, the presence of a large number of fractures and faults, the block structure of the Earth's crust. The latter is undoubtedly associated with the processes of the rising of deep-seated matter and its intrusion into the Earth's crust along the rift and transform faults. The available data indicate that on the flanks of the Mid-Atlantic Ridge the measured values of heat flow vary from 0.4 to 2.5 μcal cm^2 s^{-1}, and in the rift valley from 0.3 to 8.2 μcal cm^2 s^{-1} [182]. Investigations on the polygon in the area of the Atlantis Fracture Zone have shown that the highest heat flow values (6.9 and 7.6 μcal cm^2 s^{-1}) were observed at the intersection of the rift valley with a transverse trench, where on the slopes serpentinized peridotites were found by dredging. A considerably lower heat flow (from 0.8 to 3.1 μcal cm^2 s^{-1}) is observed on rift crests made up of tholeiite basalts [78]. Submarine observations performed within the framework of the 'FAMOUS' Project in the area south-west of the Azores have established that there is a normal heat flow on the bottom of the rift valley, and that the enhanced values are associated with the feet of its slopes, where fissures are of greater dimensions [159].

The heat flow observed in the rift zone of the Mid-Atlantic Ridge is difficult to explain on the strength of the available data on the heat conduction of rocks constituting the crust and the upper mantle. There must obviously exist a highly intensive mechanism of heat transfer from the interior of the Earth to the ocean floor surface, probably by means of convective flow in the mantle. Under the rift zones the convective flow is assumed to go upward and then on reaching the base of the lithosphere, it diverges in both directions. This accounts for the rise of the asthenospheric diapir and the spreading of lithospheric plates [82, 84, 182]. A large dispersion of measured heat flow values is explained by a random distribution of the observation points relative to the network of faults and fissures where the intrusions of deep-seated matter and heat discharge are taking place. With increasing age of the lithospheric plate the permeability of its surface layer must decrease owing to the accumulation of sediments and surficial basalt effusions. This seems to be the explanation of the general decrease in the average heat flows from the axis of the Mid-Atlantic Ridge towards its flanks, as well as the decrease in the dispersion of observed values.

CHAPTER 4

Seismicity, Volcanism and Faults of the Atlantic Ocean Floor

Seismicity, volcanism and faulting are geological phenomena with one common feature. They are all confined to weak zones in the Earth's crust where neotectonic movements and dislocations are taking place and, for this reason, should be examined as mutually interrelated.

1. SEISMICITY

Earthquakes are known to occur with the release of stresses in the Earth's crust and the upper mantle and are indicative of the present-day tectonic activity in a given region. Investigations of the Atlantic Ocean floor seismicity, as well as of that of oceanic regions in general have started comparatively recently. In the first half of our century the information on the distribution of earthquake epicentres within the Atlantic was still slight, but even this information made it possible to distinguish three main seismically active zones: the Mid-Oceanic, the Mediterranean and the Marginal Pacific which includes the Antilles and the South Antilles island arcs. In recent years the increased number of seismic stations, improvement of equipment and a greater accuracy in determining the coordinates of epicentres have made it possible to plot new seismic maps and to study the main features of earthquake foci [106, 217, 218]. The interrelation has been ascertained between the seismicity and the processes of the spreading of lithospheric plates in the zones of mid-oceanic ridges and their absorbtion in the zones of island arcs [166]. Based on the analysis of seismic data and their comparison with tectonic environment, a seismotectonic map of the Atlantic Ocean (Figure 27) has been plotted and the principal regularities in the distribution of earthquake epicentres among the morphotectonic zones have been studied [63].

As already noted, almost all earthquake epicentres in the Atlantic Ocean are confined to the Mid-Atlantic Ridge and the island arcs. Many of them are located in the Mediterranean Seismic Belt lying beyond the ocean proper. According to our calculations more than half (59%) of all the earthquake epicentres recorded in the ocean, with a magnitude above 5.0 (on the Richter scale), are concentrated in the Mid-Atlantic Ridge zone. The Antilles and

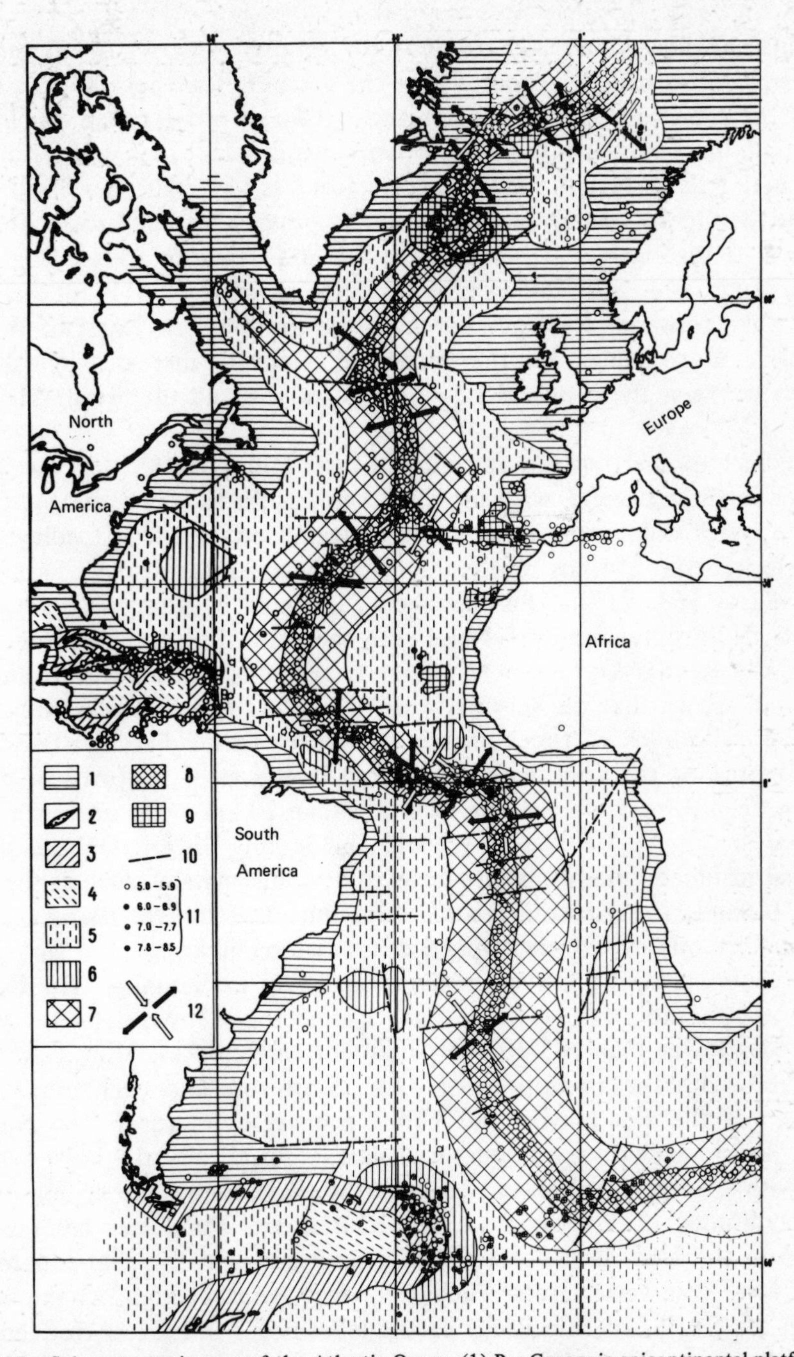

Fig. 27. Seismo-tectonic map of the Atlantic Ocean. (1) Pre-Cenozoic epicontinental platforms; (2) deep-sea trenches; (3) Cenozoic folded structures; (4) submerged median massifs (basins); (5) oceanic platforms; (6) arch-and-block rises; (7) the Mid-Oceanic Ridge flanks; (8) rift zone; (9) volcanic massifs; (10) major fractures; (11) earthquake epicentres and their magnitudes (on Richter scale); (12) axes of stresses in earthquake foci [70].

the South Antilles island arcs account for almost a quarter of all the recorded epicentres (23%). About 4% of the epicentres are located in a relatively short zone between the Azores and the Strait of Gibraltar which is a connecting link between the Mid-Atlantic Ridge and the Mediterranean Alpine Belt. The rest of the earthquake epicentres (14%) are dispersed over the Atlantic Ocean, mainly concentrated in certain specific areas. Among these can be distinguished the zones of the Mid-Labrador Ridge, the Cape Verde Islands, the Scandinavian continental margins, the Pyrenees, the zones to the south of Newfoundland, at the south-eastern coast of Brazil, the Walvis Ridge zone. Another zone of relatively increased seismicity stretches between the South Sandwiches island arc and the South-Atlantic Ridge.

Almost all the earthquake epicentres of the Mid-Atlantic Ridge are concentrated in the rift zone, with most of them, as far as can be judged from the accuracy of their coordinates determination, confined to rift valleys and transverse trenches of transform faults placed between the displaced sections of rift valleys [63, 217]. This has been confirmed by direct seismological observations on polygons, performed with the help of bottom seismographs by R/V *Akademik Kurchatov*. Comparative analysis of microearthquake records has shown that the seismicity of rift valleys and transverse trenches is tens and hundreds of times greater than that of the adjacent rift ridges.

The depths of rift zone earthquake foci in the great majority of cases do not exceed 30–35 km and are even less than 20 km in the northern part of the ocean on the Reykjanes, Iceland and Mohns Ridges. Only in some individual instances, mainly in the areas of volcanic massifs, such as the Jan Mayen, Iceland, the Azores, Tristan da Cunha and Bouvet Island, earthquakes are sometimes recorded with depths reaching 60 km.

Differences are also observed in the intensity of earthquakes. The Reykjanes, Iceland and Mohns Ridges are characterized by relatively weak (magnitudes lower than 6.5) though frequent earthquakes. More powerful earthquakes with the amplitudes of 6.5–7.5 are rare and occur mainly in the zones of the Jan Mayen and Gibbs transverse fractures. The North-Atlantic Ridge, including its equatorial part, is seismically the most active. Along with a large number of weak earthquakes numerous strong ones, with magnitudes of 7.0–8.5, are recorded there. The latter are mainly localized in the area of the Azores volcanic massif and the near-equatorial zone of large transverse faults. Their frequency is rather high. At the same time, on the South-Atlantic Ridge earthquakes occur much less frequently, but a rather large percentage of these are strong earthquakes with a magnitude of 7.0–8.5. The above facts can obviously be explained by the individual parts of the Mid-Atlantic Ridge undergoing different stages of seismotectonic development. The youngest is the northern part of the ridge, including the Norwegian-Greenland Basin. The central part of the ridge is at a later stage of development, and the southern part is the most mature.

Investigations of the mechanisms of earthquake foci in the rift zone of the Mid-Atlantic Ridge performed by different authors [70, 217] have provided evidence of the existence of the ocean floor spreading process. It has been established that the stresses developing in the foci are arranged on the horizontal plane with tensile stresses directed perpendicular to the ridge axis and compressive forces – along the axis. The release of stresses manifests itself mainly as strike-slip faults and the direction of slip always coincides with that of the displacement of rift structures along the transverse fault.

Another important characteristic of the Mid-Atlantic Ridge seismicity are the so-called earthquake series consisting of a sequence of almost continuously repeating tremors without any main strong tremor. These are regarded as reflecting the process of magmatic rocks intrusion or volcanic eruptions. According to available observations, such earthquake series are confined to the rift zone of the ridge, particularly to the regions of large volcanic structures, and are hardly ever encountered in the zones of transverse fractures [166].

In the Caribbean Sea region the zone of seismicity, branching off from the main Pacific Marginal Belt, passes along the Cayman Trench, then along the Greater and Lesser Antilles island arc, the Puerto Rico Trench, and then turns along the northern coast of Venezuela and Colombia. The maximum of seismic activity is associated with the Puerto Rico Trench and the Lesser Antilles island arc. Most of the epicentres of earthquakes with the magnitude of 5.0–7.0 are located there. The frequency of earthquakes is high. Strong earthquakes (magnitudes of 7.0–8.5) occur much more rarely. These are mainly recorded in the regions of Hispaniola, the Cayman Trench and the northern coast of Venezuela, although some individual strong earthquakes have also been recorded in the Lesser Antilles island arc area [218]. All this indicates that the eastern, frontal part of the Antilles island arc is at an early stage of seismotectonic development, whereas the sublatitudinal branches of the arc have reached a more mature stage.

The distribution of earthquake hypocentres in the Antilles island arc zone is indicative of the existence of an inclined focal plane extending from the Puerto Rico Trench axis and its continuation in the south, going under the ridge of the island arc. Along the trench and the Barbados ridge only shallow-focus earthquakes, with depths of hypocentres not exceeding 60–70 km, are recorded, while near to the southern shores of Hispaniola and Puerto Rico and along the internal slope of the Lesser Antilles island arc the depths of the majority of recorded earthquakes reach 150–200 km. At the same time, in the region of the Cayman Trench a different picture is observed. Practically all the earthqukes here, irrespective of the depth of their hypocentres, are concentrated along one zone, which may indicate the vertical position of the Cayman Trench focal plane [218].

A similar distribution of earthquake epi- and hypocentres can be seen along the South Antilles island arc. The highest seismicity occurs in the South Sandwich Trench and the adjacent arc of volcanic islands. Almost all the earthquakes in the zone of the deep-sea trench have shallow foci with depths of hypocentres reaching 60 km. In the direction towards the island arc ridge the depths of hypocentres increase, reaching 100–200 km and more under the internal slope of the arc. The number of earthquakes decreases sharply on the sublatitudinal branches of the Scotia Ridge, but rather strong ones with magnitudes of up to 6.5–7.0 predominate among them [106].

Investigations of earthquake mechanisms in the zones of island arcs and deep-sea trenches show that tensile stresses here are directed along the strike of the structures and compressive forces across it [70]. This confirms the assumption that the inclined focal planes along which earthquake hypocentres are located represent deep faults or the interfaces between the underthrusting oceanic plate and the overthrusting subcontinental crust of the island arc. The subsidence of the oceanic plate generates stresses whose release results in the earthquakes observed here. According to seismological data, the rate of underthrust in the zone of the South Sandwich island arc, e.g., is 3 cm yr^{-1} [166].

Vast areas of the ocean floor, including the oceanic basins, the flanks of the Mid-Atlantic Ridge and the inclined planes of continental rises, are seismically stable regions. Within their limits, with very few exceptions, practically no earthquakes of greater magnitude than 5 are ever recorded. This indicates, on the one hand, a calm geotectonic environment and, on the other hand, the rigidity of the lithosphere in the above regions. Only in some of the above-mentioned regions, relatively rare microearthquakes are recorded, caused by the manifestation of modern geotectonic movements inherited, in their turn, from more ancient ones. Of greatest interest are the epicentres of earthquakes in the zone of the Mid-Labrador Ridge – the echoes of former geotectonic processes, and those located along the continental margins, indicative of the continuing processes of their submergence and the formation of foredeeps.

2. VOLCANISM AND ITS ROLE IN THE FORMATION OF OCEAN FLOOR RELIEF

Volcanic processes play a very important role in the formation of the Earth's oceanic crust and the ocean floor relief. As already shown, the second layer of the oceanic crust consists of volcanogenic rocks, mainly basalts. This layer forms the basement of all oceanic morphostructures and outcrops on the ocean floor surface in the form of block-and-ridge relief of the Mid-Atlantic Ridge, as well as of numerous seamounts and, less frequently, volcanic islands. That is why all the variety of the forms of submarine relief whose development is radically affected by volcanic processes can be subdivided into several groups: (1) volcanic seamounts and islands; (2) volcanic island arcs; (3) the Mid-Atlantic Ridge; (4) oceanic arch-and block rises, swells and ridges (Figure 28). This sequence, among other things, reflects the extent of the participation of volcanic, tectonic and exogenetic factors in their formation. The role of volcanism, e.g., gradually decreases from seamounts and islands to arch-and-block rises while that of tectonics and sedimentation increases [66].

Volcanic Seamounts and Islands

There is a great number of seamounts on the Atlantic Ocean floor. Except for some mountains of block structure, located within the confines of continental margins and representing klippen, all the others are of volcanic origin. This is borne out both by the data of geological and geophysical investigations of the seamounts themselves and by the fact that all the oceanic islands, representing the summits of large seamounts raised above the ocean level, consist of volcanic rocks and some of them are active volcanoes.

The great majority of seamounts have in plan a rounded or an elliptical shape, steep and slightly dissected slopes (gradients of up to 20° and more), and pointed summits. There are crater-like depressions on some summits. Large seamounts rising close to the ocean surface have truncated, flat tops (guyots) lying at different depths from 40 to 400 m. Among the latter are, e.g., Rosemary, Anton Dorn, Atlantis, Plato, Cruiser, Great Meteor, Discovery, Vittoria, Davis, and other seamounts [151, 227]. The flat surface of the tops of these mountains seems to have resulted from abrasive – accumulative weathering, while the difference in depths has been caused by their nonuniform submergence during the Neogene-Quaternary time with a simultaneous rise of the ocean level. That the ocean levels were previously lower is indicated, in particular, by the occurrence of submarine shore terraces that have been found at the edges of some seamount summits.

The number of volcanic seamounts of the height of 1 km and more discovered on the Atlantic Ocean floor amounts to about 950. Moreover, there exist numerous seamounts of the height of less than 1 km, the true number of which is still impossible to ascertain. Since the number of seamounts regularly increases with a decrease in their height, we have calculated that on the ocean floor there should be no less than 3000 volcanic seamounts

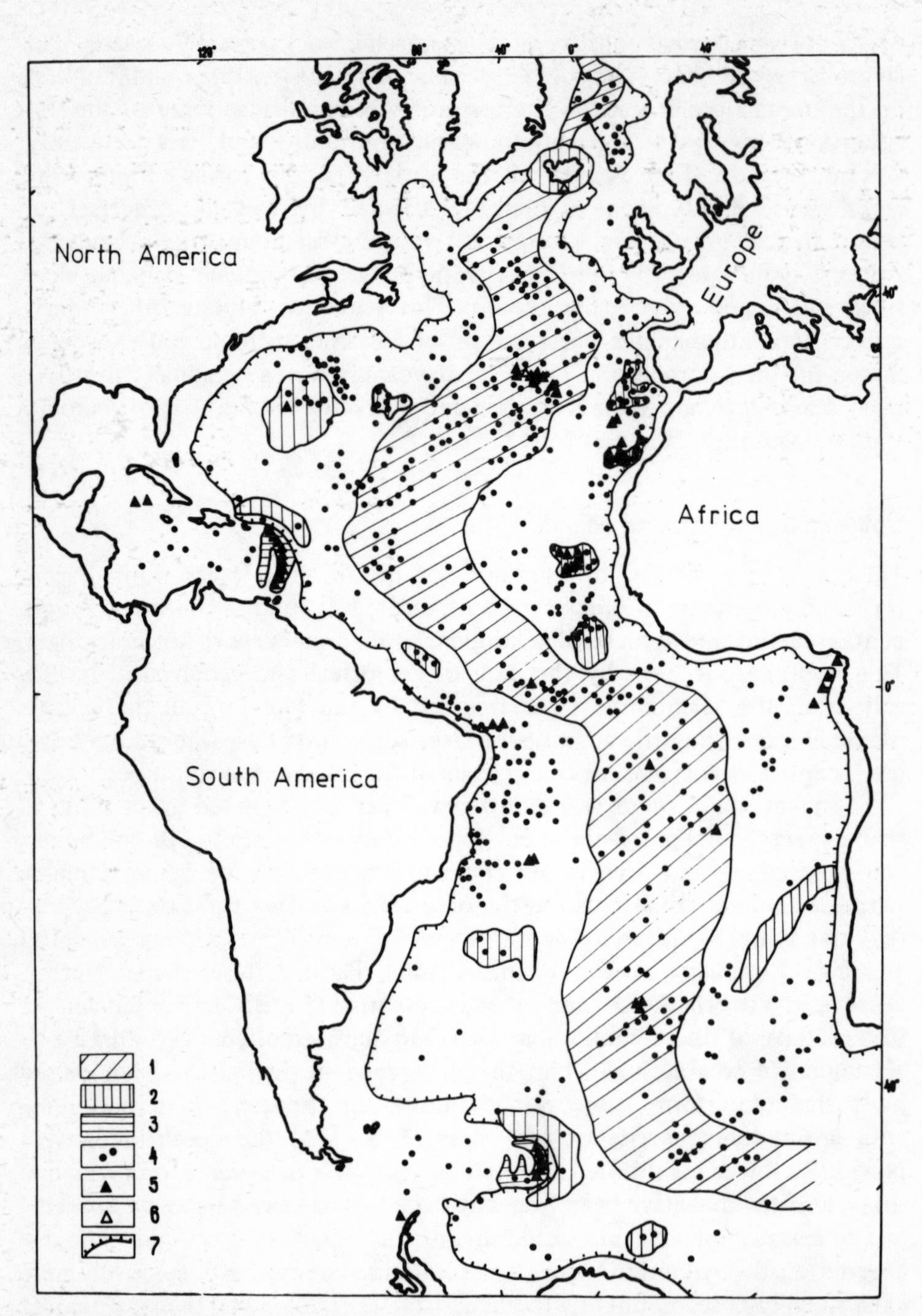

Fig. 28. Schematic map of volcanic and volcano-tectonic morphostructures and relief forms of the Atlantic Ocean. (1) the Mid-Oceanic Ridge; (2) arch-and-block rises and swells; (3) volcanic island arcs; (4) volcanic seamounts; (5) volcanic islands; (6) newly-formed volcanic islands; (7) boundary between the ocean bed and the continental margins and transition zones.

of heights ranging from 0.5 to 1 km; the total number of all seamounts will then reach almost 4000.

Certain laws govern the distribution of volcanic seamounts [62]. Within the boundaries of the Mid-Atlantic Ridge are mainly located low seamounts with a height of up to 2 km. Most of them are situated along the zones of transverse faults or in the areas of local uplifts such as the Palmer Ridge or the Azores volcanic massif. Large seamounts are rare on the Mid-Atlantic Ridge. Only to the south of the Azores on the eastern flank of the ridge is there a group of large seamounts – Atlantis, Cruiser, Great Meteor, and others. On the contrary, a considerable number of large seamounts more than 3–4 km high, as well as a multitude of small ones, are encountered on the floor of oceanic basins and the inclined plains of continental rises. In the light of the plate tectonics concept this is associated with the increasing thickness and age of the spreading lithospheric plates [233]. The positioning of small seamounts is on the whole chaotic, but large ones are mainly located along certain lines apparently demarcating the zones of deep faults. Among these, e.g., are: the New England Seamount Chain, a group of seamounts in the area of the Azores–Cape St Vincent Ridge, a chain of volcanic islands and seamounts passing along the Cameroun fault, latitudinal chains of volcanic islands and seamounts at the eastern coast of Brazil and others. Besides these, there are concentrations of seamounts and volcanic islands within the limits of arch-and-block uplifts on the ocean floor, such as the Bermuda, the Seara, the Corner, the Sierra Leone Rises, and the Rio Grande Plateau and the archipelagoes of the Canary and the Cape Verde Islands.

Our geological and geophysical knowledge of seamounts is still insufficient. Specimens of bedrocks have been obtained from a small number of seamounts, such as the Caryn, the Great Meteor, the Gorringe, the Minia, the Swallow, the Confederation, the Bald, and others [66, 130, 161]. Out of the available geophysical data of great importance are those of magnetic and gravimetric surveys. They indicate that volcanic seamounts are distinguished by clearly defined magnetic and gravitational anomalies. Three types of seamounts are in this case identified, varying in the structure of anomalous geophysical fields and confined to the basic ocean floor morphostructures. On the Mid-Atlantic Ridge the mountains are characterized by small positive free-air anomalies (up to +50 mGal) and moderate magnetic anomalies (up to 500 gammas) difficult to distinguish against the background of banded linear anomalies because of poor development of volcanic roots. On the ocean bed the seamounts are distinguished by considerable free-air anomalies (up to +300 mGal), weak Bouguer anomalies (up to +30 mGal) and strong magnetic anomalies (up to 2000 gammas), mainly associated with well developed volcanic roots. Characteristic anomalous fields are observed on the mountains of volcanic island arcs, where negative Bouguer anomalies (up to −60 mGal), positive free-air anomalies (up to +100 mGal) and strong magnetic anomalies (up to 1000 gammas) have been registered. Negative values of Bouguer anomalies are caused by the pipes of volcanic piles in island arcs being, as a rule, filled with pyroclastic material of reduced density [18].

According to the available data, it can be regarded as an established fact that volcanic seamounts on the ocean bed and the Mid-Atlantic Ridge are mainly composed of tholeiite and subalkalic basalts, while the upper parts of large seamounts, as well as volcanic islands, are usually represented by alkalic basalts, trachytes and tuffs. On some seamounts, located in the zones of transverse faults (e.g., the Gorringe Seamount to the west of the Strait of Gibraltar or the small island of São Paulo on the Mid-Atlantic Ridge), ultrabasic rocks are found: serpentinized peridotites and serpentinites. The summits and the slopes of seamounts, except for steep scarps, are usually covered with a thin layer of sediments, mostly foraminiferal oozes and aleurites, which in a number of cases happen to be cemented under the action of heat and hydrothermal solutions. The feet of seamounts and volcanic islands within the ocean bed and on continental rises are usually buried under a sedimentary cover whose thickness often exceeds 1 km.

It is quite obvious that volcanic seamounts and islands originate from the central eruption type and are confined to the weakened zones of the Earth's crust expressed as planetary deep-seated faults and differently oriented regional faults. The formation of submarine volcanoes on the ocean floor (in deep-water conditions) probably starts with the outflow of tholeiite basalts and the appearance of forms of the shield volcano type. The explosive process at great depths, because of the hydrostatic pressure, seems to manifest itself only slightly, which is indicated by the absence or a very small quantity of pyroclastic products in eruptive rocks of deep-sea seamounts. Deep-water conditions are more favourable for the extrusive process. That is why submarine extrusions, combined with lava outflow, produce huge shield volcanoes and cones several thousands of metres high on the ocean floor. The primary forms of newly appearing conic structures of lava extrusions have been found during the submarine investigations of the rift valley performed within the "FAMOUS" Project [159]. In the course of the subsequent growth of submarine volcanoes the 'primitive' tholeiite lava outflows are replaced by those of alkalic basalts. The causes of this are as yet not clear. The widely accepted opinion is that alkalic olivine basalts are derivatives of tholeiite basalts and that seamounts, in the course of their growth, create themselves the conditions in the form of additional chambers for crystallizational differentiation of material [130].

Seamount basalts are usually subjected to deuteric alterations whose intensity generally increases in the direction away from the Mid-Atlantic

Ridge axis toward continental margins. Seamounts in the rift zone (e.g., the Minia Seamount in the Gibbs Fracture Zone) are composed of fresh and only slightly changed tholeiite basalts with a small participation of peraluminous basalts and even, rarely, of subalkalic basalts. As the distance from the ridge axis increases (e.g., the Bald and the Swallow Seamounts located at a distance of 100 and 1000 km, respectively, from the rift valley) the role of alkalic basalts gradually increases. At the same time, the extent of changes in the rocks, in particular erosion, increases. The degree of deuteric alterations and of erosion of basalts must depend, among other factors, on the duration of their exposure on the ocean floor, i.e., their age. Hence, the above facts conform to the general scheme of the ocean floor spreading and the increasing age of basement rocks as one moves away from the Mid-Atlantic Ridge axis on both sides of it.

Oceanic islands, as well as many of the largest seamounts, represent, as a rule, not isolated volcanoes, but more complex formations consisting of a number of merged volcanoes. Trachybasalts, trachytes, phonolites occur here along with alkalic basalts, as, e.g., on Jan Mayen, Ascension, St Helena, Tristan da Cunha, Bouvet Island, and others [91, 93, 105]. Considerable inclusions of acidic magmatic rocks are noted in some places. In Iceland they account for 10% of all the effusiva. The profusion of pyroclastic material among the rocks of which volcanic islands and shallow submarine volcanoes are composed is noteworthy. This material is usually represented by ash strata and horizons of tuffs of basaltic composition. For instance, during the recent submarine eruptions in the regions of Iceland and the Azores violent outbursts of ash gave rise to a new island of Surtsey and to the Capalinhos Peninsula [43]. It should be stressed that shallow submarine explosive eruptions are distinguished, as a rule, by unusual force and frequency of explosions. The explanation of this can be seen in the formation of a great quantity of gases and steam, rapid cooling of the surface of lava under water and practical absence of hydrostatic pressure.

The degree of volcanic activity that has led to the appearance of the Atlantic Ocean seamounts and islands is quite high. According to calculations, the volume of material composing these forms of the ocean floor relief, taking into account their feet buried under the sediments (Table IV), amounts to 1.4 million km^3.

This material, however, is known to have been accumulated over a very long period of time. According to the results of age determinations of rocks on volcanic islands and some seamounts, volcanic structures from the Quaternary to the Upper Cretaceous occur on the Atlantic Ocean bed, i.e., their age is from tens and hundreds of thousands to 120 million years [105, 224, 237]. As a rule, the age of islands and apparently of seamounts becomes more ancient as one moves away from both sides of the Mid-Atlantic rift

TABLE IV
Dimensions of the Atlantic Ocean volcanic mountains

Relief forms	Number	Area (thous. km^2)	Volume (thous. km^3)
Volcanic seamounts more than 1 km high	950	670	490
Volcanic seamounts less than 1 km high	3000	750	180
All volcanic seamounts with feet buried under sediments	3950	–	1000
Oceanic volcanic islands	45	180	260
Oceanic volcanic islands with feet buried under sediments	45	–	400
Total	3995		1400

zone towards the continental margins. This is usually regarded as one of the indications of ocean floor spreading. The existing deviations from the above tendency (e.g., the age of Fernando de Noronha Island is 12 million years, whereas that of the Canary Islands is 32 million years, although, according to the spreading scheme, they should be at least 120 million years old) do not seem to disprove the ocean floor spreading hypothesis, but only indicate a later volcanic activity associated with local magmatic chambers in the moving plates of the lithosphere.

Submarine Ridges of Volcanic Island Arcs

The Lesser Antilles and the South Sandwich volcanic islands are located on submarine bow-shaped ridges which serve as connecting links between the sublatitudinal branches of the Antillean and South Antillean Ridges. These bow-shaped ridges, as already mentioned, are swells with relatively flattened, slightly convex summits and stepwise side slopes. From geophysical data [119, 134] the layer of volcanogenic rocks under them goes down to a depth of 2–3 km forming a bulge which determines their structures.

From the results of geological investigations on islands and the data of dredging on submarine slopes, it has been established that the ridges of volcanic island arcs are mainly composed of andesite lavas [97, 104]. On the Lesser Antilles Ridge the lavas vary from basalts to dacites and on the South Sandwich Ridge from dacites to liparites. The covering lavas are permeated by bedrock intrusions. On the Lesser Antilles the lavas are in some places covered by shallow-water limestones and tuffs of the Eocene-Miocene age, which indicates an interruption in volcanic activity, denudation and temporary submersion under the ocean level. At the end of the Neogene volcanic activity resumed. On the Lesser Antilles many of the volcanoes are still active, and in the South Sandwich Islands, judging by traces of eruptions, they were active not long ago – in the Pleistocene and the Holocene.

Therefore the dominant role in the formation of bow-shaped ridges of island arcs was played by volcanic processes lasting with some interruptions during the whole of the Cenozoic. Tectonic processes manifested in vertical movements, faults, and the dissection of ridges into blocks apparently only made these morphostructures more complicated. That is why their relief can be regarded as tectonovolcanic.

The Mid-Atlantic Ridge

As indicated above, the Mid-Atlantic Ridge represents a vast arch-like uplift on the ocean floor, its surface characterized by a typical block dissection. The basement of the ridge is made up of a volcanogenic layer up to 2–3 km in thickness, which forms its structure. Most widespread are tholeiite basalts composing the main body of the ridge rocks, its blocks and ranges. Subalkalic and alkalic basalts are encountered less frequently. They are found on islands and summits of large seamounts, as well as on the slopes of transverse trenches and the ridge flanks in the form of cross-cutting bodies in overlapping sediments. Peraluminous and plagioclase basalts also occur rarely and seem to compose individual domes and short potent flows. Intrusive bodies of dolerites, diabases and gabbros are found here and there. Ultrabasic rocks are even less frequent on the Mid-Atlantic Ridge and are met exclusively in the zones of transverse trenches [22, 130, 175, 202].

The above-mentioned variations in the composition of basalts on the Mid-Atlantic Ridge are evidently associated with tectonics and, in particular, with the way the deep-seated matter, out of which magma is generated, rises up. According to experiments [147], tholeiite magmas saturated with silica, originate at a smaller pressure than undersaturated alkalic magmas. The latter most probably appear at great depths and pour out onto the surface along abyssal faults. Tholeiite magma appears at lesser depths at optimal rates of rise, which is evidently actualized in the rift zone. Peraluminous basal magmas seem to originate from tholeiite and alkaline magmas through the dissolution of basic plagioclases.

The data on the morphostructure of the ridge and the composition of igneous rocks indicate the predominantly fissure type of volcanic eruptions which resulted in the formation of its basement. It is known that during subaerial fissure outflows the basalt lavas on land spread over vast areas (up to hundreds and thousands of km^2) and form flows of the plateau basalt type with approximately the same thickness. During submarine eruptions the picture is evidently different. Submarine basalt samples are usually composed of fragments of globular and pillow lavas with a well developed vitreous chilling crust 2–3 cm thick. Peripheral parts of lava pillows are composed of compact finely porous basalt, and the porosity increases perceptibly towards the pillow centres. Kidney-shaped spherical and cylindrical formations composed of concentric layers of overchilled glass are frequently found. All this indicates that basalt lavas cool rapidly during submarine eruptions, and so do not have enough time to spread over large distances. Probably they form swell-like piles going along eruptive fissures and consisting of short flows with globular jointing. New portions of lava seem to break apart the old covers, and globular fragments roll down the slopes giving rise to chaotic placers.

It is absolutely clear, however, that medium-size and large forms of the Mid-Atlantic Ridge, such as rift valleys and ranges, blocks, scarps, trenches, etc., result from differentiated tectonic movements with the formation of numerous faults in the course of the spreading of lithospheric plates away from the axis on both sides of the ridge. It can be said that volcanic processes have formed the ridge basement and that tectonic processes have formed its relief. The role of exogenous factors here is insignificant and manifests itself in the submarine erosion of the basalt bed and the accumulation of a thin intermittent cover of sediments filling in the interrange valleys on the ridge flanks. The relief of the Mid-Atlantic Ridge should, therefore, be regarded as volcano-tectonic [66].

There is a certain regularity in the increase of the age of basalts outcropping on the surface of the Mid-Atlantic Ridge on both sides of its axis. According to the determinations of the age of dredged samples by the potassium-argon method, and the results of deep-sea drilling, the age of the rift zone basalts does not usually exceed 10 million years, whereas on the ridge flanks it reaches 50 million and more years [161, 165]. In the rift valley the basalts are still younger. For instance, pillow lavas whose age is about 10 thousand years and less were found on the rift valley slopes during the works on the FAMOUS Project [148].

Dome-and-Block Elevations, Ridges and Swells

As noted above, the basement of oceanic elevations, such as the Bermuda and Rio Grande Plateaus, the Sierra Leone Rise, the Walvis Ridge, the volcanic massifs of the Canary and Cape Verde Islands, the Antillean and South Antillean Outer Swells, is represented by a layer up to 2–4 km thicker than that of the oceanic basins. Almost everywhere it lies under a sedimentary cover with a thickness of from 0.5 to 1.0 km, reaching in some places 1.5 km. Summit surfaces of dome-and-block elevations are, as a rule, even or slightly hilly, and the side slopes are made up of stepwise scarps apparently arising during the formation of these morphostructures as a result of differentiated vertical tectonic movements.

There are as yet not enough data on the composition of rocks in the basement of oceanic elevations. But invariably, when it was possible to obtain bedrock samples, they proved to be those of oceanic basalts or their derivatives [118, 160, 165]. Probably, here too volcanic eruptions of fissure type preodominated resulting in the formation of basalt covers. Their age, judging by the data of deep-sea drilling, ranges from the Late Cretaceous to the Paleogene.

It can thus be assumed that volcanic processes played the leading part in the formation of the basement of dome-and-block elevations and swells on the Atlantic Ocean bed. The development of their relief, however, has been primarily determined by tectonic causes: dome uplift of the oceanic crust and its dissection into blocks by systems of faults. The final shaping of submarine relief occurred during the Neogene and the Quaternary by way of the accumulation of a rather thick sedimentary cover.

The intensity and scope of volcanic activity during the formation of the oceanic volcanogenic basement, which extends not only on the ridges, elevations and swells, but also on the floor of basins, is indeed tremendous. Taking into account the areas of these morphostructures and the average thicknesses of volcanogenic basement we have estimated the total volume of volcanic rocks formed as a result of fissure eruptions on the ocean floor at about 105 million km³. The distribution of this mass of rocks over individual morphostructural provinces is shown in Table V. Taking into

TABLE V
Dimensions of the volcanogenic layer of the Atlantic Ocean

Relief forms	Area (mln. km^2)	Thickness (km)	Volume (mln. km^3)
The Mid-Atlantic Ridge	24	2–3	60
Oceanic dome-and-block elevations, ridges and swells	4	2–4	10
Floor of oceanic basins	30	1	30
Volcanic ridges of island arcs	0.4	2–3	1
Floor of the basins of marginal seas	4	1	4
Total	62	–	105

consideration the fact that volcanogenic basement rocks on the ocean floor date from the Late Jurassic to Quaternary, the average intensity of volcanic eruptions during the Meso-Cenozoic (about 180 million years) can be assumed to equal 0.5 km^3 yr^{-1}. Thus, the role of volcanism in the geological development of the ocean floor proves to be very significant, greatly exceeding what is known for the continents. This is another specific feature of the oceanic crust structure distinguishing it from that of the continental crust.

3. FAULTS

The Earth's crust, as shown by the results of geological and geophysical investigations, has a layered-and-block structure, and faults play an important role in its formation. They control the location and the strike of structures, relative displacements of these structures take place along them. The laws governing the distribution and the orientation of the continental crust faults have now been studied sufficiently [98]. Systems of faults of meridional, latitudinal and diagonal (north-western and north-eastern) strike, whose formation was brought about by the stresses in the Earth's crust associated with rotary forces and subcrustal processes, have been revealed.

On the Atlantic Ocean floor several systems of faults traced both on the continental margins and the ocean bed have also been found. An especially significant achievement is the discovery of numerous transversal transform faults of the Mid-Atlantic Ridge which play an important role in the formation of its structure. The data on the location and the strike of these faults are used in the plate tectonics concept to ascertain the position of opening poles and the direction of lithospheric plate movements in the course of ocean floor spreading [181].

We identified faults in the Atlantic Ocean floor, shown on the geomorphological and the morphotectonic maps, by their geomorphological, geophysical and geological characteristics [56]. The geomorphological characteristics include such data as the existence of linearly elongated steep scarps, narrow and deep trenches, and zones of submarine relief break-up. The second group of characteristics consists of the geophysical data on the Earth's crust structure, the existence of offsets and ruptures in individual crustal layers, zones of typical magnetic and gravitational anomalies, and seismicity manifestations. And, finally, the third group of characteristics comprises of geological data, such as the strike of on-land faults known to penetrate into the adjacent areas of the ocean floor, the existence of volcanic seamount chains on the ocean floor, zones of ultrabasite intrusions, outcrops of mylonitizated and brecciated rocks.

As already noted, two systems of large faults are clearly traced on the Mid-Atlantic Ridge: the longitudinal axial (rift) faults and the transversal (transform) faults. The former are morphologically fixed at narrow and deep rift valleys dissecting the arch of the ridge along its axis. They are associated with pronounced magnetic and gravity anomalies, enhanced heat flow and high seismicity, with the stresses in earthquake foci being indicative of the processes of the Earth's crust expansion [63, 234]. These faults extend along the whole system of mid-oceanic ridges and result from deep-seated processes leading to the rise of the asthenospheric diapir and the spreading of the lithospheric plates. The faults obviously appeared at the early stages of ridge development and are still active.

Transversal faults of the Mid-Atlantic Ridge within the rift zones are morphologically expressed by deep trenches along which one usually observes a displacement of the neighbouring rift structures, and their characteristic anomalous geophysical fields, to one side or the other. Zones of high seismicity moving together with the rift valleys are associated with these trenches [61, 155, 166]. Outcrops of ultrabasic rock are frequently met on the slopes of transverse trenches and the adjacent sections of rift valleys, which is indicative of the intrusion of deep-seated matter. The age of ultrabasites, as a rule, turns out to be greater than that of surrounding basalts, usually the Paleogene or even the Late Cretaceous [110]. Of very ancient age – more than two billion years – are the ultrabasites of the San Pâolo Island rocks. They are assumed to be the elevated stocks of mantle matter, and their age reflects the age of the upper mantle [195]. On the flanks of the Mid-Atlantic Ridge transverse faults are traced as series of scarps, crosswise orientated valleys or zones of submarine relief break-up with which more or less clearly defined magnetic anomalies are, as a rule, associated. These sections of transverse faults are practically aseismic, as also are the flanks of the ridge as a whole [63, 155]. The data presented obviously indicate that the transverse faults are more ancient than the rift zone and the whole of the Mid-Atlantic Ridge. They are the controlling factor in the formation of the ridge structure and are responsible for its being divided into a number of segments. The sections of transverse faults located in the rift zone happen to be involved in the process of the ocean floor spreading and, therefore, they are still tectonically active. The nature of the movement of lithospheric plates in the region of the Mid-Atlantic Ridge explains the peculiarities of the morphology of transverse faults (Figure 29). On the ridge flanks the neighbouring sections of plates on both sides of the fault move in the same direction from the ridge axis while in the rift zone they move in the opposite direction, which results in the formation of a deep and relatively wide transverse trench whose depths exceed those of the neighbouring rift valleys.

Apart from the large faults described above, on the Mid-Atlantic Ridge

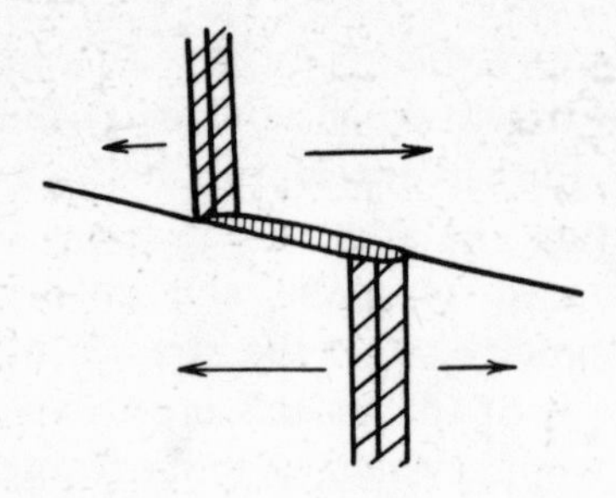

Fig. 29. Scheme of the movement of lithospheric plates in the zone of transform faults.

there is a great number of local tectonic disturbances in the form of small faults and fissures which create a distinctive block-and-range structure of the ridge. As a rule, these deformations in the sections between transverse faults are orientated along the strike of the ridge. In some places one finds dislocations with a different strike, associated with the existence of local structures diversifying the ridge. Among these can be mentioned the region of the Azores, the Palmer and Mesjatzev Ridges, and other structures.

On the floor of oceanic basins large faults are traced either along the chains of volcanic seamounts and islands or along the linear scarps bounding the block ridges or dome-and-block elevations. Among the former are: the Cornwall Fracture along the New England Seamount Chain in the North-West Atlantic Basin [224]; the system of East-Azores faults running along the Azores–Cape St Vincent Rise and along the Josephine, Gorrige, Ampere and other seamount chains [179]; the submarine extension of the Cameroun Fracture passing through the volcanic islands of Fernando-Póo, Principi, São Tome, Annobon via the chain of seamounts in the Angola Basin up to the St Helena Island [60]; a series of latitudinal faults along the chains of seamounts and volcanic islands of Fernando de Noronha, Trinidade and Martin Vaz to the east of the coast of Brazil within the Guiana and Brazilian Basins [62], and a number of other faults. Among the latter should be reckoned a series of diagonal faults along the south-eastern margin of the Bermuda Plateau [168]; the faults in the region of the Canary Islands, established from the geological observations on the islands and the geophysical works around them [111]; faults bounding the Walvis Ridge, the Sierra Leone and Seara Rises, the Rio Grande and Rockall Plateau, and others [56]. The appearance of these fractures is evidently associated with differentiated vertical tectonic movements because of which the oceanic elevations and ridges have been formed. Their age corresponds to that of the elevations and, as shown by the results of deep-sea drilling [165], does not go beyond the Late Cretaceous. On the other hand, the faults traced along volcanic seamount chains are most probably the result of shearing stresses arising from the horizontal movements of the Earth's crust. Their age can be fairly ancient, not coinciding with that of volcanic mountains, since the latter evidently arose in the already existing zones of crustal weakness.

A system of faults is traced along continental margins and island arcs of the Atlantic Ocean, associated in one way or another with the structures of continents and their submarine continuations. Steep scarps of the continental slope bounding the regions where ancient shields or rejuvenated mountain massifs come close to the coasts have undoubtedly been formed along the lines of disruptive dislocations and represent large faults or systems of faults. This is confirmed by the data of seismic profiling and other geophysical investigations performed in a number of regions [128, 142, 129, 215]. Such marginal faults formed where marked differentiated vertical movements were taking place: the rise of land and the down-warping of adjacent ocean floor sections with the formation of foredeeps. The age of these marginal faults can vary but is mostly Jurassic-Cretaceous, if one takes into account the age of sedimentary strata on the submarine margins of continents where they manifest themselves.

Another type of marginal faults is traced in the shelf regions of glacial areas – along the coasts of Newfoundland, Labrador, Greenland, Scandinavia. The fractures here separate the massifs of coastal land undergoing neotectonic uplifts from the submerging outer shelf made up of a thick sedimentary series. These faults are morphologically expressed as marginal trenches, separated from the coastal part of the shelf by steep scarps. They are relatively recent. They are considered to have been formed in the course of the neotectonic uplift of the coastal parts of dry land during the Neogene and were then renovated by glacial-isostatic movements during the Pleistocene [68, 163].

In the zones of the Antillean and South Antillean island arcs longitudinal faults are clearly traced, limiting the arc structures and manifesting themselves either as deep trenches or steep stepwise scarps. The arcs themselves are dissected by series of obliquely transverse faults into a number of blocks, which is clearly seen in the submarine relief. Within the basins located inside the arcs are also identified differently orientated, mainly latitudinal and diagonal faults which limit or separate individual structures, such as the Nicaragua Rise and the Beata Ridge in the Caribbean Sea or the medial zone in the western part of the Scotia Sea. Zones of faults going down under the island arcs (Benioff seismic zones) and outcropping on the ocean floor in the form of the deep-water Puerto-Rico and South Sandwich Trenches should be especially noted. Judging by the seismicity, these faults are tectonically active up to now, and their origin should be related to the Paleogene or even to the Late Crateceous. Sublatitudinal fractures bounding the basins of the Caribbean and Scotia Seas along the feet of the ridges of island arcs, judging by geological data, served as a kind of main lines, along which these regions were subjected to an eastward displacement during the Paleogene and the Neogene [97]. It was obviously as a result of this displacement, on the one hand, and the counterwise movement of the Atlantic Ocean floor, on the other hand, that the inclined plates of the Benioff zones were formed, along which the island arcs were thrust over the oceanic plates.

Examining the map of the Alantic Ocean faults as a whole, one notes that faults of longitudinal and transverse strike mainly occur in the zones where linear elongated structures (the Mid-Oceanic Ridge, block ridges, island arcs) are located. The azimuths of these strikes change in accordance with the general direction and the bends of the linear structures themselves. One sees, however, that in most cases fractures and their sections coincide with the general planetary network of orthogonal (meridional and latitudinal) and diagonal strikes. This indicates that the orientation of faults on the ocean floor has been caused, among other things, by the stresses associated with the action of the Earth's rotary forces.

Marginal faults limiting the continental massifs, as a rule, dissect the fold structures of ancient shields or rejuvenated mountain systems at different angles. The azimuths of strikes of these fractures for the most part have a north-eastern or north-western direction which correlates with the diagonal network of faults typical for continents.

On the ocean bed the sublatitudinal faults are the largest. They are seen in the majority of transverse transform faults of the Mid-Atlantic Ridge, a number of faults along the chains of volcanic seamounts on the floor of oceanic basins and some faults dissecting continental margins and extending into the continents. Moreover, many of the transverse faults of the Mid-Oceanic Ridge are clearly seen to be located on the continuation of the latitudinal faults of continental margins, and are obviously connected with them. In a number of cases there is geological and geophysical evidence of this [42, 99, 183]. It can be assumed that we are dealing here with planetary latitudinal faults dissecting the Earth's crust of the Atlantic Ocean from one continent to another. Together with submeridional and diagonal faults, which are seen to be of minor significance, they divide the Earth's crust into a great number of large blocks (or segments). The existence of shear deformations along the faults is indicative of the mutual displacement of these blocks in the course of the Atlantic Ocean tectonic development.

CHAPTER 5

Tectonic Movements and the Development of the Atlantic Ocean Floor Morphostructure in the Meso-Cenozoic

The morphostructural plan of the Earth's surface is known to have been formed as a result of prolonged geological development, starting from the Proterozoic when the continental nuclei originated. Subsequenly, different horizontal and vertical movements of lithospheric plates, folding, sedimentary cover formation, denudation and other processes repeatedly changed the aspect of our planet, gradually bringing it closer to that of today. The main features of the present-day morphostructural plan were formed in the Late Mesozoic and the Early Paleogene, and it was finally shaped at the neotectonic stage of development beginning from the Oligocene, i.e., approximately over the last 40 million years.

The results of prolonged development of the ocean floor have been fixed in its present-day structure and are reflected in the accompanying morphotectonic map (Figure 30). The map has been plotted on the strength of the geological, geophysical and geomorphological data described above and is based on the morphostructural principle. In the current state of knowledge on the ocean floor structure we regard this principle as the most appropriate; moreover, it reflects the fact that, as a result of peculiarities characterizing the development of the Earth's oceanic crust, its geological structure is sufficiently well expressed in the submarine relief [55]. Structures of different order are distinguished on the map. The planetary structures consist of mobile belts (geosynclinal and mid-oceanic) and stable plates (continental and oceanic platforms). Within their limits are located large morphostructures, such as ridges, swells, dome-and-block elevations, volcanic massifs, depressions and trenches. Various disruptive dislocations dissect the ocean floor into regional plates or blocks, stressing in this way the layered and block structure of the Earth's crust. The main role in their formation, as indicated by the geological and geophysical data, was played by endogenic factors: tectonic movements accompanied by faults and by volcanic processes owing to which the oceanic basement was formed.

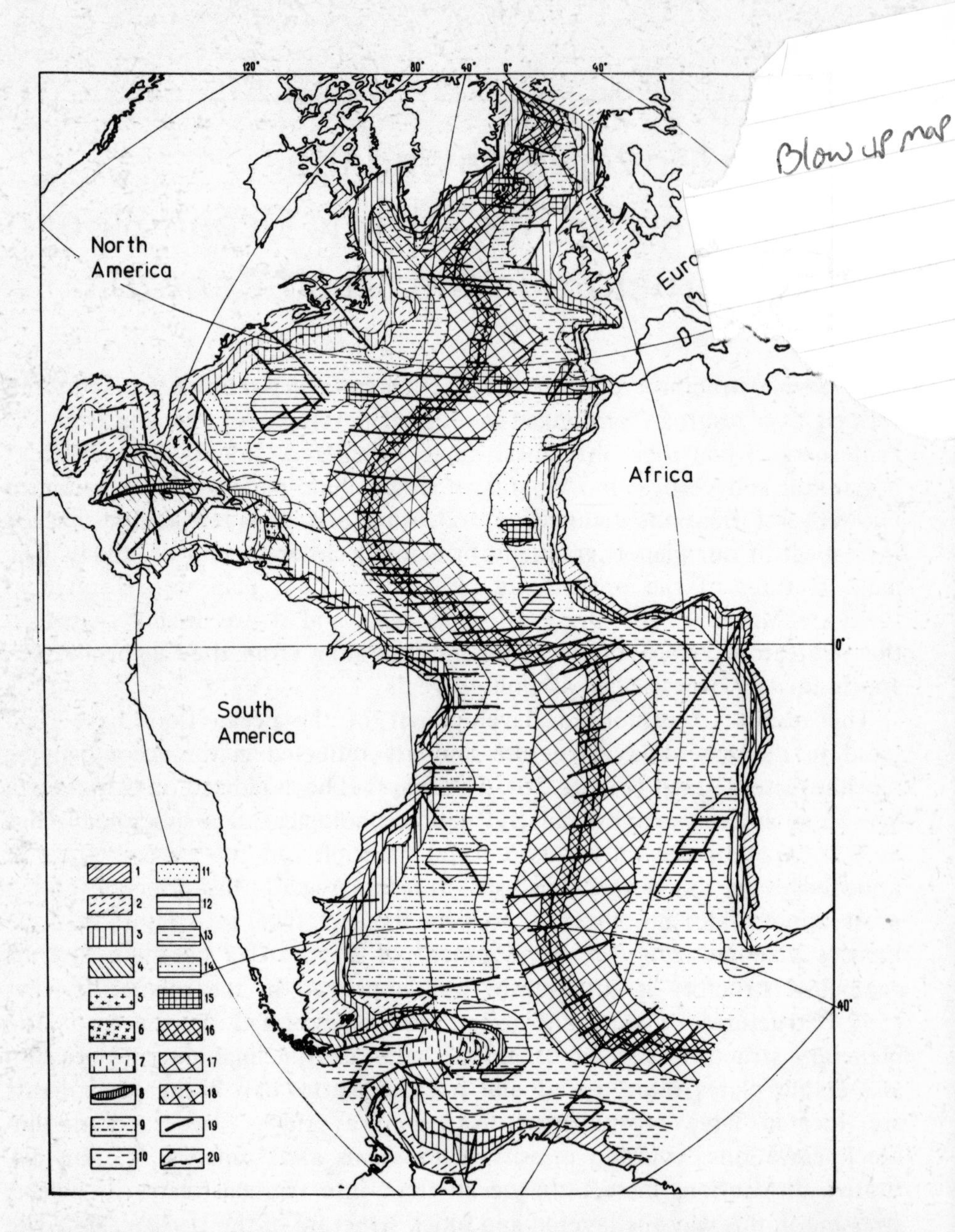

Fig. 30. Schematic morphotectonic map of the Atlantic Ocean. (1) Precambrain platforms; (2) Epi-Palaeozoic platforms; (3) foredeeps; (4) Cenozoic folded structures; (5) present-day volcanic island arcs; (6) extinct volcanic arcs; (7) submerged median massifs (basins); (8) deep-sea trenches; (9) oceanic platforms with thick sedimentary cover; (10) oceanic platforms with thin sedimentary cover; (11) accumulative ridges; (12) arch-and-block rises; (13) marginal arc swells; (14) block ridges; (15) volcanic massifs; (16) rift zone; (17) flank zones of the Mid-Oceanic Ridge; (18) buried parts of the Mid-Oceanic Ridge; (19) faults; (20) morphological boundaries between the Mid-Oceanic Ridge and the foot of continental slope.

1. HORIZONTAL TECTONIC MOVEMENTS

The information presented above indicates that both the horizontal and the vertical tectonic movements have played and are obviously still playing the decisive role in the formation of the Atlantic Ocean floor morphostructure. As yet little is known about the causes of these movements but their results are seen clearly enough. The idea of gravitational convection in the Earth's mantle and the enormous stresses arising in the Earth's crust as a result of it seems to be the most acceptable in explaining the observed phenomena [4, 84]. The process of gravitational differentiation apparently leads to the formation of the Earth's core and the initiation of convective currents in the mantle, and these, in their turn, induce the movement of lithospheric plates over the asthenosphere surface. According to this model, the oceanic crust is formed in rift zones owing to hydration of rising mantle rocks, and the continental crust – in geosynclinal regions owing to dehydration of oceanic crust in the course of its subsidence into the upper mantle along the Benioff zones.

It should be noted that there also exist other points of view on the trends of the horizontal movements in the Atlantic Ocean region. The Mid-Atlantic Ridge is supposed to have been formed by ocean floor compression at the final stages of tectonic development after the completion of a spreading cycle [44, 76]. Some authors think that compressive processes were taking place in the Atlantic Ocean region during the whole of the Meso-Cenozoic [72]. However, the absence of traces of such compression within the limits of the Mid-Atlantic Ridge in the form of folded or overthrust structures make us disagree with the above assumptions.

According to the concept we accept, in the Atlantic Ocean region the major and most pronounced horizontal movements are those of lithospheric plates in the direction away from both sides of the axial rift fault. The results of these movements are seen in the symmetry of the Mid-Oceanic Ridge morphostructure and the banded magnetic field, in the orientation of stresses in earthquake foci of the rift zone, in the distribution of the thickness of sedimentary cover and the age of basal deposits and rocks of the volcanic foundation, in the resemblance between the outlines of continental margins on both sides of the ocean, and in other facts. The kinematics of these movements has been elaborated comprehensively enough within the concept of new global (plate) tectonics on the basis of the strike of banded magnetic anomalies, rift zones and transform faults [181, 184]. Quantitative calculations of the rates of lithospheric plates spreading are mainly based upon the data on the location and strike of banded magnetic anomalies [37, 124, 204].

On the strength of the above-mentioned data it has been established that the rate of the Atlantic Ocean floor spreading during the Cenozoic was 1–2 cm yr^{-1}, with the highest rate (about 2 cm yr^{-1}) noted in the northern part of the Southern Atlantic, decreasing to the south and the north. In the southern part of the South Atlantic and in the North Atlantic up to the Gibbs Fracture Zone the rate of ocean floor spreading amounts to 1.5 cm yr^{-1}, decreasing to 1 cm yr^{-1} and less to the north of the Gibbs Fracture Zone and in the Norwegian-Greenland Basin. For the Mesozoic history of ocean floor development a higher rate of spreading, reaching 3–4 cm yr^{-1}, has been established, which is associated with a rather fast expansion of the oceanic crust after the initial stage of tectonic opening of the Atlantic and its transformation into an oceanic basin [183, 204, 235].

Since the movement of lithospheric plates takes place not on a plane but on a spherical surface of the Earth their relative displacement is actually brought about by the rotation of these plates relative to one another around a certain centre – pole of rotation. It so happens that the direction of movements on both sides of the Mid-Oceanic Ridge axis goes along the zones of transverse transform faults. That is why these faults, as well as the data on the rates of ocean floor spreading, can serve as indicators in calculating the location of the rotation poles of lithospheric plates [181]. However, because of insufficient accuracy of the initial data determination (of fault azimuths and spreading rates) and the effect of diverse local factors, the results of determining the rotation poles of individual lithospheric plates by different authors prove to be different. Nevertheless, notwithstanding a rather high dispersion of the values obtained, in general it can be regarded as an established fact that the movements of lithospheric plates in the Atlantic Ocean at the last stage of development took place by way of rotation around three poles. For the South Atlantic the pole of rotation is situated in the area of the southern part of Greenland (about 67° N and 39° W). For the North Atlantic it is situated slightly further to the north-east (about 70° N and 33° W), and for the region to the North of the Gibbs Fracture Zone, including also the Norwegian Sea, it is removed far to the east (about 65° N and 139° E). It should also be stressed that at the earlier stages of ocean floor development the plates, as calculations show, were moving in a slightly different way, relative to other rotation poles, which for the South and North Atlantic were situated to the south of the present-day ones [37, 181, 183, 204].

What is important for us is not so much the location of rotation poles as the difference, established from these data, in the movements of lithospheric plates in different regions of the ocean caused by the existence of several large plates here. In the Atlantic Ocean region one can distinguish the Eurasian and the African Plates separated by the Mediterranean Mobile Belt, the North American and the South American Plates separated by the

Caribbean Transition Zone and the small Caribbean Plate located within its limits, as well as the Antarctic Plate separated from the South American Plate by the South Antillean Transition Zone. The demarcation line between the Eurasian and the African Plates in the oceanic part (to the west of the Strait of Gibraltar) is marked by the East Azores Fracture Zone, and that between the South American and the Antarctic Plates in the area to the east of the South Sandwich Islands goes along a fault zone indicated by a seismicity belt [166, 181]. As to the demarcation line in the oceanic part between the North American and the South American Plates it can most probably be drawn along the Barracuda Fracture although it is practically aseismic.

Taking into account the above considerations and proceeding from the data on the occurrence of banded magnetic anomalies and the results of deep-sea drilling [165, 205], we have plotted a schematic map of the horizontal movements of the Atlantic Ocean floor in the form of isochrones corresponding to the position of the rear edges of lithospheric plates that were being spread over a certain time period (Figure 31). The 10-million year isochrone approximately coincides with magnetic anomaly 5 and corresponds to the outer boundary of the rift zone of the Mid-Oceanic Mobile Belt. It indicates the extent of ocean floor spreading at the last stage of neotectonic development from Late Miocene to Recent. The 40-million year isochrone, corresponding to magnetic anomaly 15, approximately coincides with the boundary between the upper and the lower parts of the Mid-Oceanic Ridge flanks. It indicates the extent of ocean floor spreading during the whole of the neotectonic stage of its development starting from the Late Eocene. This isochrone, as seen on the map, is traced to the south of Iceland along the whole of the ocean up to the boundary with the Indian Ocean, but to the north of Iceland it is absent. According to the latest data, the region of the Iceland Plateau represents a relic of the old continental massif that submerged under the ocean level. The Mid-Oceanic Ridge has broken a road for itself through this massif, represented by the Central Graben of Iceland and the Iceland Ridge, which are modern rift zones formed during the Neogene-Quaternary time. In the Norwegian Basin, however, the Aegir Ridge has been found buried under the sediments, and it can be regarded as an extinct rift zone. This rift zone was probably active in the Paleogene, then its development stopped, and the axis of ocean floor spreading shifted into the area of the Iceland Plateau. The connection between this old rift zone and the rift zone of the Mid-Atlantic Ridge is a complicated question. Judging by the deep-seated structure of the Faeroes-Iceland Rise [27], it can be assumed that these rift structures were connected along the zone where the rise joins the Iceland volcanic massif through a system of transform faults. On the other hand, it is also supposed

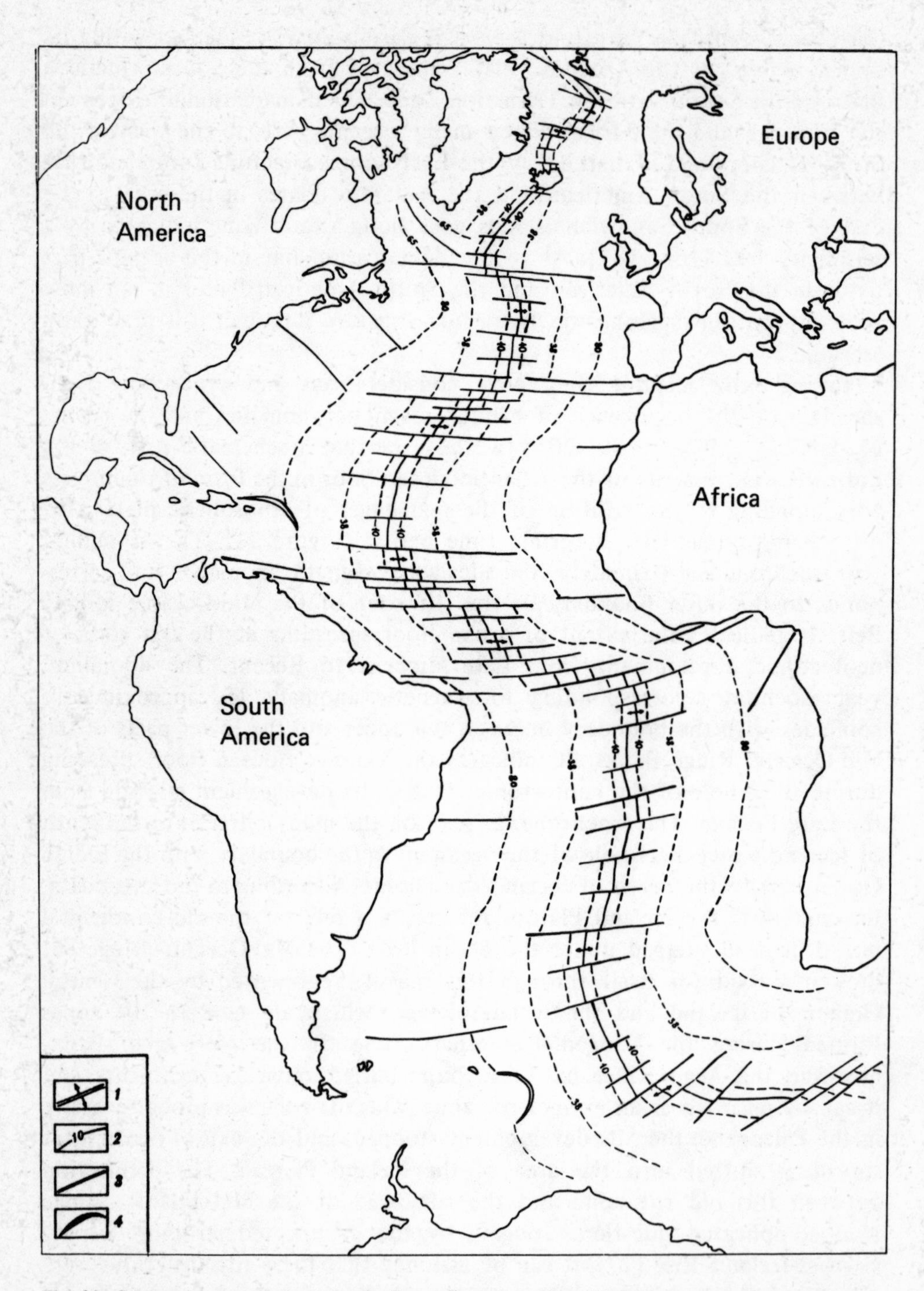

Fig. 31. Schematic map of the Atlantic Ocean floor spreading. (1) axial fracture; (2) isochrones of ocean floor spreading, million years; (3) main faults; (4) Benioff zones.

that the northern part of the rift zone of the Mid-Atlantic Ridge extended in the Paleogene into the region of the Labrador Basin where the now buried Mid-Labrador Ridge is situated. As shown by the results of magnetic surveys and the data of deep-sea drilling [178], the rift zone here was active from approximately 70 to 45 million years ago. Then its development stopped, and the axis of spreading shifted into the area of the modern Reykjanes Ridge.

The 80 million years isochrone, corresponding to magnetic anomaly 32 and pertaining to the middle of the Upper Cretaceous, is traced in the Atlantic Ocean only to the south of the Gibbs Fracture Zone. It passes approximately along the present-day boundaries between hilly and flat abyssal plains of oceanic basins. At the latitude of 50° N and in the Equatorial Atlantic this isochrone almost reaches the continental margins, indicating the limits of ocean floor spreading over the above-mentioned period of time. To the north of the Gibbs Fracture Zone the 80-million year isochrone is not traced as this region underwent spreading later, starting from the Paleogene. The peripheral parts of the ocean bed in the North and South Atlantic, located beyond the boundaries of the Upper Cretaceous isochrone, spread in a somewhat different fashion and, probably, independently of one another, since at that time the Equatorial Atlantic was closed up because the continental margins of South America and Africa were joined together. A detailed study of some of the areas in these regions makes it possible to outline a scheme of horizontal movements and the periodicity of magnetic field reversals during the Cretaceous [235]. This, however, refers to the time when the Atlantic Ocean did not exist in the present-day sense of the word, and there were only comparatively small isolated basins. We shall, therefore, not go into details of the horizontal tectonic movements of that period.

A complicated picture of horizontal movements reveals itself in the Mexican-Caribbean Transition Zone. From the data of geological and geophysical studies it is assumed that North and South America were closer to each other in the Triassis. The Yucatan and the Nicaragua-Honduras continental blocks were squeezed in between them so that there was a united dry land massif here [99, 145]. Then in the Jurassic, at the time when the Atlantic Ocean began to open up, North America started moving away from South America, and these blocks underwent rotation, gradually approaching their present-day state. During the Cretaceous, the Antillean geosyncline was formed, and the Caribbean plate enclosed by it expanded to the north and the south because of the crust extension and also underwent a certain movement to the east, thrusting over the oceanic plate which moved under it along the Benioff zone. This is confirmed by the latitudinal faults passing along the northern and southern branches of the Antilles arc and by the Puerto Rico Trench with the Barbados Ridge located in a deep trough. This pattern of horizontal movements was obviously also preserved in the Paleogene, but in the Neogene vertical tectonic movements became predominant in the Caribbean Sea.

Similar complicated horizontal movements took place in the South Antillean Transition Zone. In the first stages of the development of this region the predominant feature was the extension of the Earth's crust and the formation of the Drake Passage resulting from the withdrawal of South America from Antarctica. The Scotia Sea plate then started moving eastwards, which brought about the formation of a sharply curved arc of the South Antillean Islands. These movements are also fixed in the crust structure and the banded magnetic anomalies of the Drake Passage and in the sublatitudinal shear faults of the Scotia Sea. In the last stages of the development of this region a convergence of South America and Antarctica probably took place.

The presence of horizontal tectonic movements in the Mid-Atlantic Ridge rift zone is also indicated by some geological observations. Firstly, these are the data on the occurrence of dikes penetrating the strata of basalts in Iceland. These could quite obviously be formed under the conditions of the Earth's crust extension. The most recent investigations and calculations based on the data of the Soviet Geodynamic Expedition [87] show that the rate of crust spreading in the Holocene was 1 cm yr^{-1}, and that during the Pliocene-Quaternary time the possible spreading of Iceland rift zone could have reached scores of kilometers. Secondly, in the 'FAMOUS' Project the data of direct geological observations from submarines have been obtained in the rift valley at 36–37° N [148]. Here on the area of about 6 km^2 more than 400 fissures have been recorded, mainly orientated along the rift valley, with the width from hairline cracks near the axis of the valley up to 10 km at its edges. The depths of the these fissures range from 10 to 100 m. This is an unquestionable indication of the progressing extension of the Earth's crust on both sides of the ridge axis.

Local horizontal tectonic movements are observed in the zones of transverse faults. Many, though not all, of these are transform faults, i.e., they connect the neighbouring sections of rift structures. They appeared even before the riftogenesis and served as controlling factors in the generation of spreading axes. In the sections of faults situated between the displaced rift valleys lithospheric plates move in opposite directions causing shearing deformations, while beyond the boundaries of rift valleys the neighbouring sections of plates move along the faults in the same direction, so no shearing occurs. In some instances, however, as indicated by geological and geophysical data, tectonic shifts are observed all along the fault zone. Individual local shifts are also established along some faults dissecting the continental margins but the magnitude of these is in general not great.

2. VERTICAL TECTONIC MOVEMENTS

Apart from horizontal movements an important role in the formation of ocean floor structure is played by vertical movements. It is known [98] that one of the main methods of identifying them is the analysis of the thickness and the facies of sedimentary rock strata. In combination with the data on the ocean floor relief this method makes it possible to obtain information on the trend and scope of diverse vertical movements of the Earth's oceanic crust [58]. Based on the above information concerning the ocean floor geomorphology, the structure and thickness of sedimentary cover, the faults and volcanic structures, as well as the data obtained by deep-sea drilling, we have plotted a map of the vertical movements of the Atlantic Ocean floor in the Meso-Cenozoic (Figure 32).

Because of the basic differences between the continental and the oceanic crust, calculations of vertical movements have been performed separately for the ocean bed and the continental margins including the transition zones. The boundary between them, as already noted, is formed by the foot of continental slopes or that of the outer scarp of marginal plateaus, as well as by deep-sea trenches of island arcs.

The configuration of the ocean bed has been reproduced based on the model of lithospheric plates spreading away from both sides of the Mid-Atlantic Ridge axis. According to this model and the calculations performed [83, 211], a practically continuous rise of asthenospheric matter, its intrusion through rift faults, interaction with oceanic water, cooling and crystallization were taking place in the rift zone of the ridge during the Meso-Cenozoic, at the same time as ocean floor spreading. Crystallization of silicates is known to be accompanied by an increase in their density. As the Mid-Atlantic Ridge, according to gravimetric data, is on the whole isostatically balanced, it should be assumed that as a result of the above processes the axial zone of the ridge must be situated at a certain more or less constant level, while the ridge flanks and the floors of basins gradually submerge as one moves away from the axis. This assumption is in general confirmed by deep-sea drilling data [165]. In boreholes drilled in abyssal basins of the ocean at the basement of a sedimentary sequence carbonate deposits are invariably found, in contrast to the upper layer of sediments where practically noncalcareous red deep-water clays predominate. This points to the fact that the bottom of modern oceanic basins was previously situated above the critical depth of carbonate dissolution, which in the Atlantic Ocean is generally 4800–5300 m [50]. At the same time, the deep-water facial composition of these carbonate deposits shows that they were deposited at a sufficiently large depth amounting to thousands of metres. Consequently, in the course of spreading the ocean floor gradually submerged and passed the critical depth

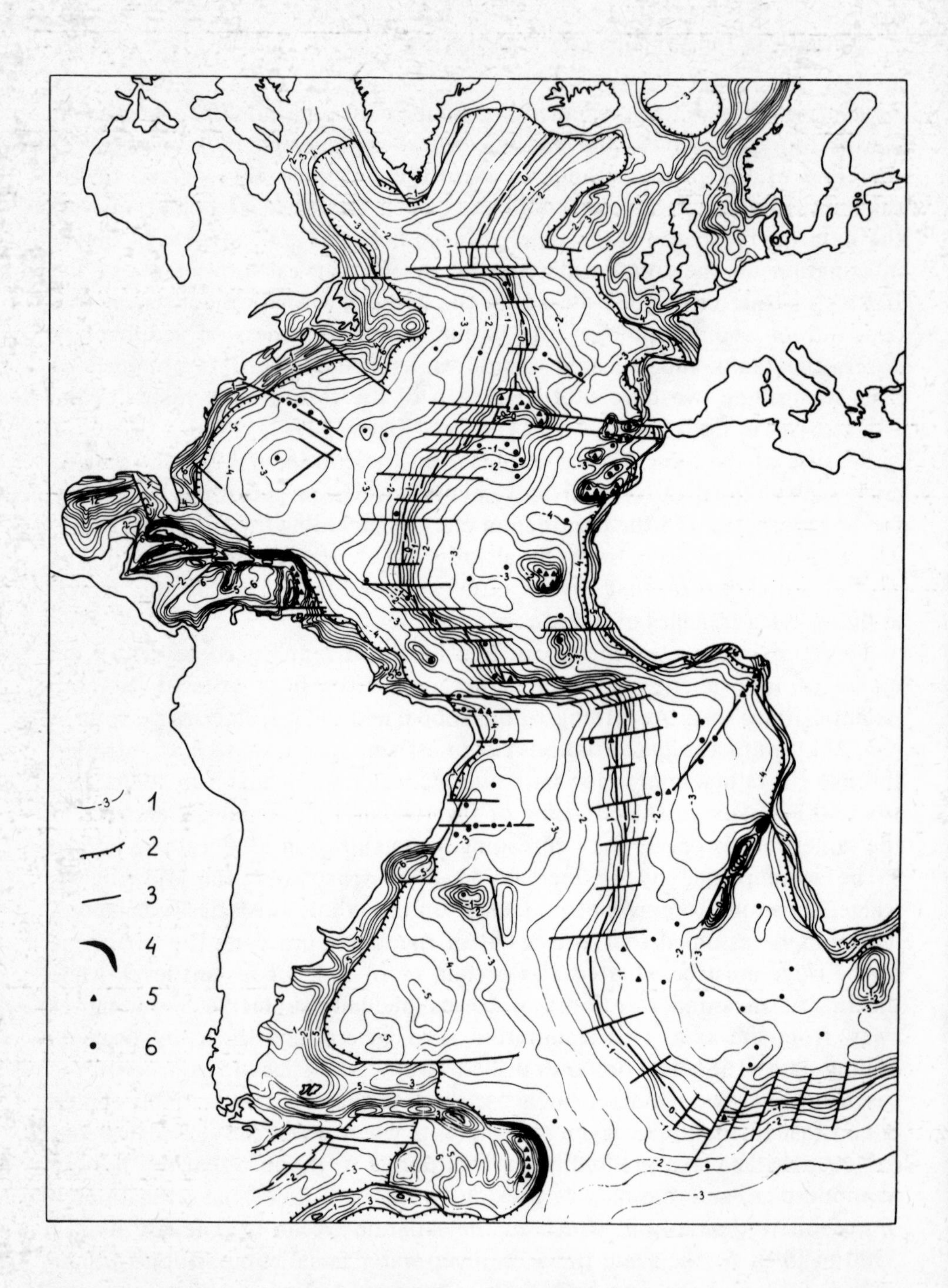

Fig. 32. Schematic map of vertical movements of the Atlantic Ocean floor. (1) isoamplitudes of vertical movements in km; (2) boundary between continental margins and ocean bed; (3) main faults; (4) deep-sea trenches; (5) volcanic islands; (6) volcanic seamounts.

level. In was obviously in this way that the Mid-Oceanic Ridge and the oceanic basins were developed as forms of submarine relief. The average depth at which the Mid-Atlantic Ridge crest is located has been taken as the initial reference level of the amplitudes of ocean bed vertical movements. On most of it this depth amounts to about 2.5 km, but in the vicinity of Iceland, because of a general decrease in ocean depth, it does not exceed 1 km. The amplitudes of vertical movement shown on the accompanying map have been calculated from a bathymetric map and a diagram of sedimentary cover thicknesses. Since the total thickness of sediments down to the oceanic basement surface has been taken for calculations, the resulting amplitudes of movements, naturally, account for the whole time of the deposition of these sediments.

For continental margins and transition zones the configurations have been reproduced based on the fact established in numerous areas that here, during the Meso-Cenozoic, submersion of the continental crust and accumulation of sedimentary rock masses of varying thickness was taking place, and this was practically continuous, though unequal in extent [57, 208]. Ancient deposits on continental margins, as a rule, have a shallow-water facial composition which is indicative of their formation in a littoral zone. Therefore, the reference level to determine the extent of vertical movements on continental margins should correspond to the ocean level, which in the Mesozoic was in fact different, but by not more than several hundred metres, from the present-day level [47]. In calculating the extent of movements on continental margins in order to correlate them with ocean bed movements only Meso-Cenozoic deposits were taken into account. In many regions (at the eastern coast of the U.S.A., to the west of Great Britain, at the western Coast of Africa, etc.) these deposits form an undisturbed sedimentary cover occurring on dislocated rocks of the Upper Paleozoic, which makes them clearly distinguishable on seismic sections. In other regions (e.g., the North Sea, the Gulf of Mexico) the sedimentary cover also includes Paleozoic deposits. These had to be subtracted from the total thickness of sediments. In the Caribbean Sea the sedimentary cover occurs on a volcanogenic basement, and it is not older than the Late Cretaceous. The surface relief of this basement has been determined by vertical movements, and the sedimentary cover only simulates it. A similar picture is observed in the Scotia Sea, but its eastern part is associated with a transition zone and the western part with the ocean bed.

Analysis of the map leads to certain conclusions on the direction and the amplitudes of the Atlantic Ocean floor vertical movements [58]. In different regions submersion characteristic for continental margins has manifested itself in a different way. On the margins of Pre-Cambrian shields (Canadian-Greenland, Brazilian, Sierra Leone) the magnitude of subsidence is not high, the basement on the shelf lies close to the floor surface. On the margins of Palaeozoic platforms, on the contrary, the folded basement is observed to submerge to a considerable depth, which is particularly noticeable in the regions of continental foredeeps. The magnitudes of vertical movements here reach 3–6 km and more, which corresponds to an average subsidence rate of 0.02–0.04 mm yr^{-1} in the Meso-Cenozoic. In the basins of transition zones the magnitudes of subsidence are also high, reaching 5–6 km, but the history of tectonic development here is more complicated. From the data on the tectonics of island arcs and adjacent continental massifs it has been established that in the Caribbean and the South-Antillean Transition Zones in the Cenozoic there was a repeated alternation of subsidences and uplifts of various structures, and the formation of the modern morphostructural plan started in the Eocene [99]. Therefore, taking into account the above period of time (40–50 million years), average rates of vertical movements at the neotectonic stage of development here must reach 0.12–0.15 mm yr^{-1}. For instance, a slight uplift (up to 0.5 km) is noted for the region of the Azores, and a slight submersion (less than 0.5 km) – for certain sections of the rift zone in the South Atlantic. On both sides of the rift zone the ocean floor is subsiding as the spreading proceeds, and the magnitudes of subsidence regularly increase as one approaches the continental margins. In foredeeps the submersion reaches 4–5 km and more. Local relative uplifts of individual dome-and-block morphostructures are distinguished against the background of the general ocean floor subsidence.

Along with the spreading processes, the oceanic crust also underwent vertical movements associated with them, mainly submersion. The Mid-Atlantic Ridge rift zone composed of continuously renewed material has probably remained at almost the same level during the whole cycle of ocean floor spreading, and was only subjected to small regional fluctuations. Volcanic structures and oceanic crust faults make this picture even more complicated. Taking into account the age of the oldest deposits in the sedimentary cover at the oceanic bed periphery, one can estimate the average rates of oceanic crust subsidence in the Meso-Cenozoic at 0.03–0.04 mm yr^{-1}. Of about the same order are the rates of the relative uplift (to be more precise, lagging behind in subsidence) for the dome-and-block elevations on the ocean bed. It is important to emphasize that as one approaches the Mid-Atlantic Ridge axis the magnitudes of ocean floor subsidence decrease with a simultaneous decrease in the age of basement rocks. This results in the rates of vertical movements remaining approximately the same throughout the ocean bed.

Apart from the large-scale vertical movements described above, which govern the formation of the basic ocean floor morphostructures, regional and local movements are also noted, manifesting themselves in certain individual submarine relief forms. On the continental shelves, against the background of general submersion, local relative uplifts and subsidences are revealed, expressed in depth fluctuations of the flooded coeval shore terraces and the bending of the outer margin of the shelf [33]. On steep scarps of the continental slope in the areas of differentiated tectonic movements, fault surfaces and banks with clearly defined morphological features are formed. Fault banks are also very clearly seen on the slopes of deep-sea trenches, which have been studied well on the polygons in the Puerto Rico and the Cayman Trenches [65].

On the ocean bed local tectonic movements of the Earth's crust manifest themselves on the Mid-Atlantic Ridge, in fault zones and on the fault banks of dome-and-block elevations. It can be said that the block-and-range relief of the Mid-Atlantic Ridge is essentially determined by the existence of systems of faults and fissures along which the local vertical movements of the Earth's crust occur in the course of ocean floor spreading. The newest movements in the rift zone in the form of upthrow and downthrow faults are almost everywhere revealed from the data of detailed geological surveys, echo-sounding, seismic profiling and direct observations from submarine apparatuses.

Subsidence is thus the predominant type of vertical movement on the Atlantic Ocean floor. It occurs on both the ocean bed and the continental margins. This conclusion is also supported by the data of geomorphological observations made on the shores of continents and islands, by the information on the distribution and occurrence of submerged shore terraces on continental and insular shelves, and by the data on the existence of submerged abrasive-accumulative plains and the table-like summits of seamounts [49, 62, 151]. Such a clear-cut trend in vertical movements indicates that in the Cenozoic the ocean extended its dimensions not only by floor spreading but also by absorbing the continental margins in the course of their submersion. On the other hand, the above data indicate that vertical movements are not predominant in ocean floor formation. Vertical movements are interconnected with horizontal movements, and these are obviously interdependent. The role of each of these types of tectonic movements is quite definite. Horizontal movements determine the distribution and mutual location of the morphostructures of the ocean floor and the surrounding continents, and vertical movements shape the basic forms of relief, their height and dissection.

Horizontal and vertical tectonic movements also play an important role in the sedimentary cover formation on the ocean floor. Their direction and amplitude explain the presently observed peculiarities in the thickness distribution and the age of sedimentary rock sequences. According to the model described above, during the Meso-Cenozoic a constant rise of asthenospheric matter in the rift zone resulted in an approximately constant level of the crest of the Mid-Atlantic Ridge from which the ocean floor is spreading (Figure 33). In the course of spreading the lithospheric plates are

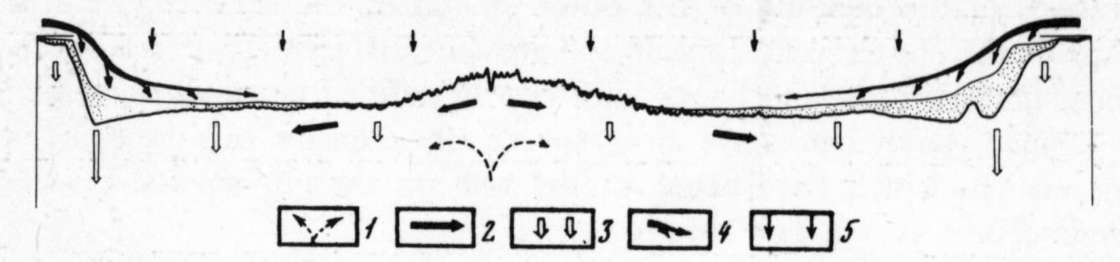

Fig. 33. Model of the ocean floor horizontal and vertical movements and the sedimentary cover formation. (1) deep-seated movements in the upper mantle; (2) horizontal tectonic movements; (3) vertical tectonic movements; (4) scattering of terrigenous material; (5) deposition of biogenic material.

consolidated and submerge. On continental margins, because of a possible withdrawal of subcrustal matter, the Earth's crust undergoes submersion, which also involves the periphery of oceanic platforms, and this results in the formation of foredeeps. Terrigenous material removed from land is gradually deposited on continental margins and in oceanic basins. In the course of ocean floor spreading the distance over which this material is scattered gradually increases, and younger sediments occupy ever-increasing areas. Hardly any terrigenous sediments penetrate into the Mid-Atlantic Ridge area, and it is mainly the biogenic carbonaceous sediments precipitating from ocean water masses which are deposited here. These are also deposited on the floor of oceanic basins, but there they are diluted with terrigenous material, and at the depths below 4800–5300 m carbonaceous material dissolves. The sedimentary cover thickness on the ocean floor, naturally, increases on both sides of the ridge axis because of the duration of sedimentation processes becoming longer as one moves away from the axial rift fault.

3. THE MAIN STAGES IN THE DEVELOPMENT OF OCEAN FLOOR MORPHOSTRUCTURE IN THE MESO-CENOZOIC

As already mentioned, when discussing the history of development of the modern Atlantic Ocean floor morphostructure we proceed from the concept of plate tectonics, regarded as the principal theoretical concept, within the framework of which the formation of peripheral parts of the ocean is considered on the basis of the notions constituting the 'oceanization' hypothesis. Analysis of the facts and data presented above leads to the conclusion that three main stages should be distinguished in the development of the ocean floor: (1) the stage of the opening-up of the ocean (Late Jurassic–Early Cretaceous); (2) the stage of the formation of the ocean floor basic morphostructures (Late Cretaceous – Early Paleogene); (3) the neotectonic stage of the final formation of the modern morphostructural plan (Late Paleogene–Quaternary). These stages are illustrated by the paleomorphostructural schemes (Figure 34(a)–(c)) we have drawn up on the basis of the above-mentioned schemes of horizontal and vertical floor movements taking into account the reconstructions made by other authors of positions occupied by the continents at different stages of geological development. The coastline on the schemes is shown more or less arbitrarily with a dashed line since its true position at those geological epochs has not been established precisely.

The Stage of the Ocean Floor Opening-up

As indicated by the available information on the geological structure of continental margins, the paleomagnetic data on the positions of continents, the data on the distribution of banded magnetic anomalies, the rates of ocean floor spreading and the age of rocks in the oceanic basement [28, 165, 177, 183, 204], in the middle of the Mesozoic, i.e. 200 million years ago, the Atlantic Ocean did not exist, and the continents of North and South America, Europe, Africa, and apparently Antarctica, were in close proximity to one another. This is confirmed by the well-known similarity of the outlines of continental margins and the geological structure of the confronting parts of the continents on both sides of the ocean [117]. Numerous authors draw the boundaries along which the continents converged coinciding with the outer margins of shelves or the feet of continental slopes. We are, however, inclined to support X. Le Pichon who thought that the continents converged at the boundaries of the zones of quiet magnetic field [181]. Indeed, the zones of quite magnetic field extend along the feet of continental slopes and on the whole coincide with continental foredeeps characterized by an increased sedimentary cover thickness. These regions have undergone a considerable submersion throughout the whole history of ocean development. With the help of seismic methods in a number of places, traces of a 'granite' layer have been found wedging out under the sediments. It can therefore be assumed that the marginal parts of joined continents subsequently submerged and subjected to 'oceanization' were situated along the zones of continental foredeeps. The continents of North and South America also lay in close proximity to each other, and what is now the Caribbean Sea was probably dry land [99].

During the Jurassic, because of a possible change in the plan of convective currents in the Earth's mantle [84] or some other processes, the unified primeval continent began to break up into a number of plates which started moving away from the axial fault on both sides of it. Thus occurred the opening-up of the Atlantic Ocean; the first areas to open up were evidently some comparatively small ones in the North Atlantic and then in the South Atlantic, in consequence of which elongated narrow basins of the modern Red Sea type were formed. On the strength of magnetic data and the results of deep-sea drilling the opening-up of the North Atlantic is assumed to have started 180 million years ago and that of the South Atlantic 140 million years ago [183, 204]. Along the axis of the new basins active rift zones were formed, where the generation of a new oceanic crust began, while the peripheral parts of the basins were formed by the submersion of the margins of continents moving away from one another. The developing rift zones also underwent submersion with the spreading of lithospheric plates, and, most probably, were initially graben-like structures. The planetary systems of fractures dissecting the previously joined continents continued their development on the newly formed oceanic crust, serving as the controlling factor in the rift zone formation and later in the formation of the whole system of the Mid-Oceanic Ridge. In the second half of this stage (the Late Cretaceous) when the unified Atlantic Ocean was formed out of separate basins the rift zone reached the depth of about 2.5 km and was then stabilized at this bathymetric level determined by the isostatic equilibrium between the rising asthenospheric diapir and the generated oceanic crust.

During the opening of the ocean the rates of lithospheric plates spreading were at first probably not high – less than 1 cm yr^{-1}. This is indicated by the analysis of magnetic data and comparison with the history of development of the Red Sea. The initial narrow basins gradually became wider and longer. In the middle of the Cretaceous the northern basin probably extended from 15° to 52° N, and the southern basin from 20° to 60° S. It is difficult to say anything definite about the expansion of the ocean floor between Africa and Antarctica, but one can suppose that this region too started to open up at the end of the Jurassic or the beginning of the Cretaceous. In the second half of the Cretaceous the Equatorial Atlantic began to spread. A graben-like trench was formed here dissected by numerous transverse faults into a series of segments displaced relative to one another. This trench connected the northern and the southern basins, which resulted in the formation of a unified Atlantic Ocean whose shape and dimensions still differed from the modern ones. The rates of ocean floor spreading had markedly increased by that time and, judging by magnetic data, amounted to 2–3 cm yr^{-1} and more. The width of the ocean by the end of the Cretaceous probably ranged from several hundred km to 2000–2500 km.

The formation of the Antillean and South Antillean Transition Zones started at the same time as the ocean generation. In the middle of the Jurassic, as a result of possible northward movement with a simultaneous westward displacement of North America, two sublatitudinal troughs were formed in the Antillean region: in the area of the Gulf of Mexico and the Colombian and the Venezuelan Basins. The Yucatan and the Honduras continental blocks moved southwards and westwards undergoing a certain degree of clockwise rotation [99, 145]. In the Cretaceous a free connection opened up between the Gulf of Mexico and the newly-born Atlantic Ocean, and the central part of the Gulf was transformed into a deep-sea trough. A second sublatitudinal trough connected the Atlantic Ocean with the Pacific. The Caribbean plate was formed here with the suboceanic type of the Earth's crust. Along its periphery appeared the Antillean-Caribbean geosyncline whose northern and southern branches penetrated into the continental margins of North and South America, and the eastern branch developed on the oceanic crust. On the western and eastern margins of the Caribbean plate large zones of faults took shape along which the Benioff zones were later formed with the Central American (in the Pacific) and the Puerto Rico (in the Atlantic) deep-sea trenches. This was accompanied by the Caribbean plate moving somewhat eastwards. A similar picture of ocean floor development seems also to have taken shape for the South Antillean Transition Zone, with the opening of the Scotia Sea probably resulting from the northward movement of South America away from the relatively stationary Antarctica. The Scotia Sea plate formed in this way also moved eastwards, and that is why a Benioff zone developed in front of the arc of the future South Sandwich Islands.

The Stage of the Formation of the Basic Ocean Floor Morphostructures

By the end of the Cretaceous, about 75–80 million years ago, as a result of the spreading of lithospheric plates and the submersion of continental margins the Atlantic Ocean had appeared and taken shape (Figure 34, a). Its further development went by way of expansion and the formation of basic morphostructures, although some of them had already appeared at the initial stage.

The rift zone where the new oceanic crust was generated continued to remain at the depth level of about 2.5 km. As the lithospheric plates spread away from the axis on both sides of it they gradually consolidated and submerged. This resulted in the appearance of the Mid-Atlantic Ridge as a form of relief which in its further development only expanded owing to ocean floor spreading and underwent local vertical movements because of the uneven rise of mantle diapir in different sections of the ridge. Taking into account the fact that Cretaceous deposits in the Atlantic Ocean are mainly represented by carbonaceous sediments, it should be assumed that the depths of oceanic basins at that time did not exceed 4500–4800 m. In

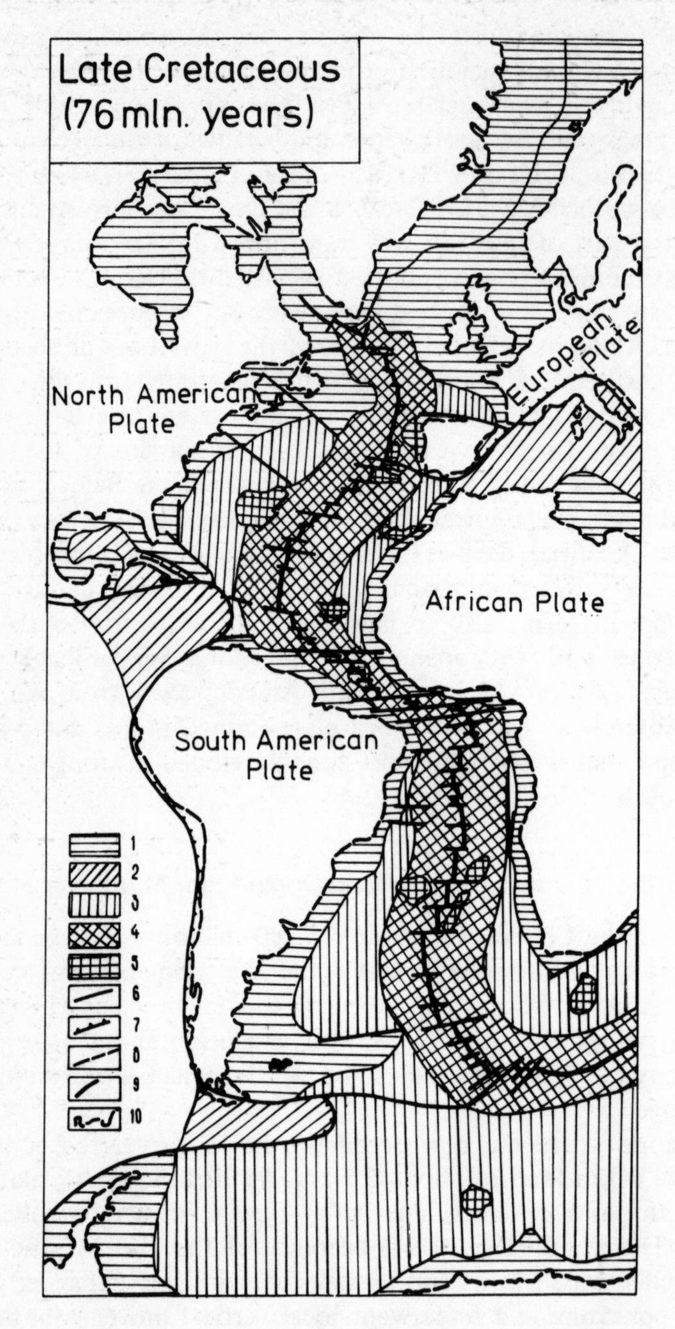

Fig. 34(a)

Figs. 34(a)–(c). Paleomorphostructural schemes of the Atlantic Ocean for the Late Cretaceous (a), Late Eocene (b) and Late Miocene (c). (1) continental margins; (2) transition zones; (3) oceanic basins; (4) the Mid-Oceanic Ridge; (5) oceanic rises; (6) faults; (7) Benioff zones; (8) island arcs; (9) rift faults; (10) approximate contours of continental plates.

the Paleogene, however, judging by a rather sharp decrease in the carbonization of deposits, the floor of the largest oceanic basins (North-West Atlantic, Brazilian, Argentine, Canary, Angola) had already passed the level of the critical depth of carbonates dissolution and reached the depth level of more than 5000 m. The sea floor spreading rates in the North Atlantic gradually decreased during the Late Cretaceous and the Early Paleogene from 2.5 to 1.5 cm yr^{-1}. In the Equatorial and South Atlantic the spreading rates were more stable, and consequently by the end of the stage the South Atlantic had begun to spread faster than the North Atlantic [204, 212]. As a result of this, during the Late Cretaceous and the Early Paleogene the Atlantic Ocean expanded by about 2000–2500 km and acquired outlines close to those of today.

The foredeeps already formed at the initial stage along the periphery of the ocean continued to submerge intensively involving the marginal parts of continents in this process. The foredeeps were filled up by sedimentary material removed from dry land, and their basement was subjected to partial 'oceanization' caused by the deep-seated processes. By the beginning of the Paleogene the foredeeps had already been almost completely filled with sediments, and the formation of huge accumulative series started in these places. Blocks of continental crust separated in a number of places from continental margins served as barriers arresting the sedimentary material, as a result of which large steps began to form here and later turned into marginal plateaus. On the submerging continental margins epicontinental platforms in the form of a sedimentary wedge bounded on the outside by a flexure or a system of faults were created by the accumulation of terrigenous sediments. These platforms served as the basis for the future continental shelves.

On the ocean bed, along with the processes of lithospheric plates spreading, regional and local vertical crustal movements took place accompanied in some places by faulting which led to the formation of individual morphostructures, such as the Bermuda Plateau, the Seara Rise, the Rio Grande Plateau, the Walvis Ridge, the socles of volcanic massifs of the Canary and the Cape Verde Islands. Origination of these structures can be assumed to have already occurred in the rift zone as a result of deep-seated processes and active fissure volcanism in much the same way as it took place at the neotectonic stage in the region of the Azores. Vertical dome-and-block uplifts and outflows of large masses of basal lavas obviously resulted in the formation of plateau-like morphostructures in the rift zone, which were then displaced together with the spreading plates into the oceanic basins where a sedimentary cover started accumulating over them. At the end of the Cretaceous and the beginning of the Paleogene all the main positive morphostructures of the floor of the Atlantic Ocean basins seem to have already been formed since, judging by the data of deep-sea drilling, they underlie distrubed carbonaceous sediments of the Upper Cretaceous and the Paleogene.

At the end of the Cretaceous riftogenesis started developing to the north of the Gibbs Fracture Zone. At first it penetrated into the area between Greenland and Labrador where the Mid-Labrador Ridge was formed along the spreading axis. Active development of the rift zone of this ridge, judging by magnetic data and those of deep-sea drilling [178, 204], continued up to about 45–47 million years ago, after that the ridge began to die out, submerged and was covered with a thick sequence of sediments.

About 60 million years ago the ocean floor began to spread in the region between Greenland and Europe. Before that the continental margins here were in close proximity to one another, with the 'continental bridge' evidently including the Faeroes – Rockall Massif and the Iceland Plateau. True, the latter most probably was of smaller dimensions, which subsequently increased because of crust spreading in the zone of the Iceland Ridge. To the south of Iceland the rate of ocean floor spreading was about 2 cm yr^{-1}, whereas in the Norwegian – Greenland Basin it was not more than 1 cm yr^{-1} [170]. The rift zone passed along the axial line between the South-East Greenland and the Rockall Plateau as well as along the now buried Aegir Ridge in the Norwegian Basin and further along the Knipovich Ridge. The connection between the rift zones of the Norwegian Basin and the North Atlantic, in our opinion, must have been brought about through a system of transform faults passing along the southern margin of Iceland and the zone where Iceland joins the Faeroes –Iceland Rise. During the Late Cretaceous and the Early Paleogene the region to the south of Iceland completely opened up over a distance of 500–600 km, and an elongated water body about 200–300 km wide – the Norwegian – Greenland Basin –was formed. Because of the spreading of the Canadian – Greenland and the West European blocks the latter started moving south-eastwards, whereas before that it moved eastwards. This accounts for a change in the nature of movements along the Azores-Gibraltar boundary between the European and the African plates. Up to the period of 60 million years ago mainly a righthanded displacement was developing here, and later a certain compression began which apparently led to the formation of upthrow faults in the oceanic crust in the form of seamount chains of the Horseshoe Rise.

At the time of ocean floor spreading in the region between Greenland and Europe the continental margins submerged, and the foredeeps and epicontinental platforms were formed. At the end of the Cretaceous the Faeroes – Rockall Massif separated from the continent, was broken up by faults and partially submerged; a block of continental crust also separated from the continent and subsequently served as the basis for the formation of the Norwegian Plateau. The submergence of the Iceland Plateau and its separation from Greenland by a foredeep formed here began in the Early Paleogene. All these differentiated vertical movements obviously caused massive outflows of plateau basalts mainly taking place in subaerial conditions and covering the area from Greenland to Scotland [6]. The older basalt sheets later submerged together with the Iceland Plateau and the Faeroes-Rockall Massif, and the new outflows in the Neogene formed the plateau basalt sheets in Iceland.

In the Caribbean Sea region the Late Cretaceous and the Early Paleogene was the time of the rapid and intensive development of the Antillean Geosyncline [99]. The period of geosynclinal subsidences was replaced by the orogenic period characterized by tangential deformations, intrusions of granitoids and the rise of the Antillean Ridge, including the Aves Ridge. The earlier basalt volcanism was replaced by andesite volcanism. The activity of movements along the Benioff zone at the eastern periphery of the ridge reached its maximum. The central parts of the Colombian and Venezuelan Basins that had been formed by the beginning of the Paleogene underwent a gradual submersion caused, probably, by the Earth's crust extension due to the northward and southward movements. As a result, in the sublatitudinal branches of the Antillean arc overthrusts were formed pointing towards the Florida – Bahama platform in the North and towards the South American platform in the South. In the Eocene the Grenada Basin was, evidently, also formed as a result of the Earth's crust extension in the rear of the Lesser Antilles volcanic arc that continued to move eastwards. On the whole it can be noted that in the Early Paleogene the entire Caribbean region was characterized by rather intensive horizontal movements along the sublatitudinal faults. The largest of these caused the appearance of the deep Cayman Trench.

A similar picture of the development of basic morphostructures can be visualized for the South Antillean Transition Zone as well. Here also, in the Late Cretaceous and the Early Paleogene, active orogenic processes took place along the previously generated geosyncline, which caused the uplift of the South Antillean (Scotia) Ridge and the submersion of the basin of the Scotia Sea. Local vertical shoves caused the appearance of block rises and ranges on the ocean floor. The general movement of the plate to the East along marginal faults led to the active development of the Benioff zone and the formation of the deep South Sandwich Trench.

The Neotectonic Stage

In this way by the end of the Eocene the basic morphostructures of the Atlantic Ocean floor had been formed, their relative positions fixed, and the ocean assumed the outlines close to those of the present, only its width and the depths of oceanic basins were less than they are today (Figure 34, b). At the neotectonic stage the submarine relief was finally shaped, the ocean floor expanded further and its basins deepened. The neotectonic movements, as a rule, were inherited from the more ancient ones, which emphasizes the successive development of the ocean floor during the Meso-Cenozoic cycle of the lithospheric plates spreading.

The rift zone to the south of the Gibbs Fracture Zone down to the boundary with the Indian Ocean remained throughout the whole stage at the previous bathymetric level of depths (about 2.5 km), undergoing only local deformations and shoves. The most substantial elevation formed at this stage is the Azores volcanic massif, representing a plateau bounded by faults, with the surface diversified by numerous volcanic islands and seamounts. The appearance of such elevated sections in the rift zone can be caused by anomalously heightened inflows of asthenospheric matter from the Earth's interior. When, however, the diapir rise activity is reduced the bathymetric level of the rift zone is observed to go down, as is the case in some sections of the South-Atlantic Ridge. Volcanic plateaus of smaller size than the Azores Plateau, created by major outflows of basalt lavas, have been found in the areas of the St Helena and Tristan da Cunha Islands.

The spreading rates of the Atlantic Ocean floor at the neotectonic stage were almost constant or slightly increased by the end of the Neogene. In the North Atlantic, as already mentioned, the spreading rates were 1.2–1.5 cm yr^{-1}, in the South Atlantic up to 2 cm yr^{-1} [64, 124, 204]. At present Africa is moving away from South America along an azimuth of 70°, and from Antarctica of 50°. Europe is moving away from North America along the azimuth of about 110°, and it should be noted that these movements are relative.

To the north of the Gibbs Fracture Zone, the depth at which the rift zone is situated decreases, as one approaches the Greenland-Britain Rise, from 2 to 1 km; this is explained by the influence of a very sharp asthenospheric diapir rise in the region of Iceland. In the Norwegian-Greenland Basin the rift zone at the beginning of the Oligocene went along the presently buried Aegir Ridge. But in the middle of the Oligocene it began to die out, and riftogenesis started developing further to the west, in the region of the Iceland Plateau, and in this way separated the portion of the plateau with the Jan Mayen Ridge from the continental margin. This happened, judging by magnetic data, about 30 million years ago [170]. The axis of spreading probably went along the eastern edge of the present-day Iceland Ridge. In the Miocene, about 10 million years ago, the spreading axis shifted still further to the west taking its present-day position. The southern continuation of this rift zone passed through Iceland and joined the rift zone of the

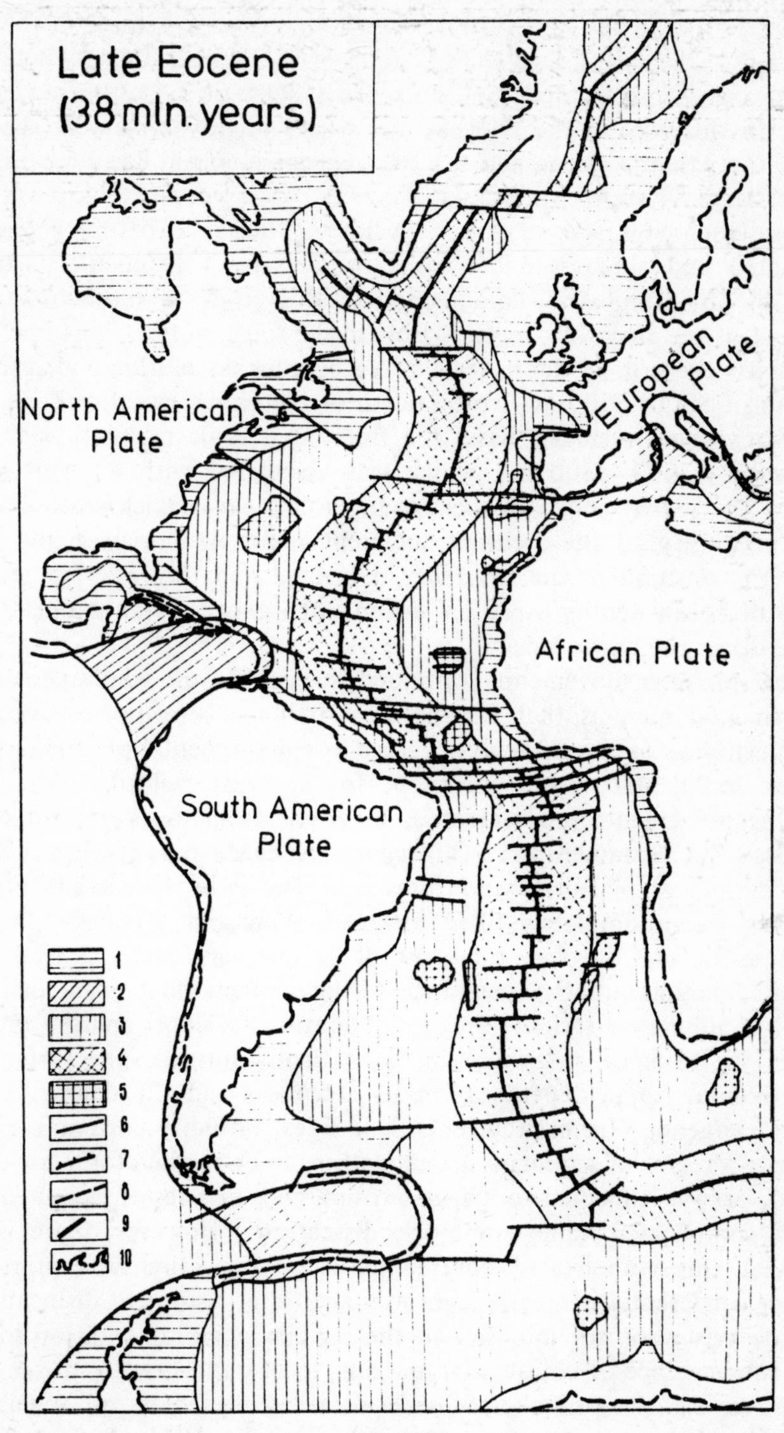
Late Eocene
(38 mln. years)
North American Plate
European Plate
African Plate
South American Plate
1
2
3
4
5
6
7
8
9
10

Fig. 34(b)

Reykjanes Ridge. In the north the Iceland Ridge rift zone, through the Jan Mayen transform fault, joined the rift zone of the Mohns Ridge, where the spreading axis is represented by a complex system of a faults arranged in echelon. In this way in the Pliocene the whole mighty system of the rift zones of the Atlantic Ocean and the Norwegian-Greenland Basin was finally created (Figure 34, c).

Lithospheric plates, spreading on both sides of the axis as before, gradually consolidated and submerged becoming overlain by a sedimentary cover. Judging by the lithological composition of this cover, at the neotectonic stage most of the floor of oceanic basins to the south of the Gibbs Fracture Zone passed the critical depth level of carbonates dissolution and reached the depth of 5000–5500 m. To the north of the Gibbs Fracture Zone and in the Norwegian-Greenland Basin the floor of oceanic basins subsided to lesser depths (3000–3800 m), which was associated with the geological youth of the region and with great sedimentary cover thicknesses caused by the proximity to the areas of sediment removal. Moreover, the high hypsometric position of the rift zone in Iceland and the resulting general decrease of ocean depths is caused, as indicated above, by an anomalously intensive diapir rise under Iceland.

Vertical tectonic movements continued throughout the whole period on the continental margins that for the most part had already been formed by the beginning of the neotectonci stage as epicontinental platforms with a varying sedimentary cover thickness. In the great majority of regions descending movements predominated, inherited from more ancient ones. This caused the accumulation of Oligocene – Neogene deposits that covered the epicontinental platforms and foredeeps. The latter, as already noted, had already been completely filled with sediments and, therefore, did not obstruct as before the free transport of sedimentary material from the continental margins onto the ocean bed. At the neotectonic stage the inclined plains of accumulative series at the feet of continental slopes and the abyssal plains on the floor of oceanic basins were finally formed. The surface of the continental slope, depending on its steepness and the degree of the effect of exogenous factors (currents, landslides, turbidity currents), either represented a region of sediments accumulation, and was, therefore, gradually flattened, or was subjected to marine erosion and, as a result, acquired the form of complicated valley-and-block dissection . However, there is no doubt that this complicated continental slope dissection was primarily caused by differentiated tectonic movements and disruptive dislocations. Such processes were not intensive in the regions where, in addition to the general submergence of continental margins, local opposite-sign movements occurred, as, e.g., in a number of sections along the coasts of Greenland, Scandinavia, Labrador, South America and Africa [97]. Differentiated

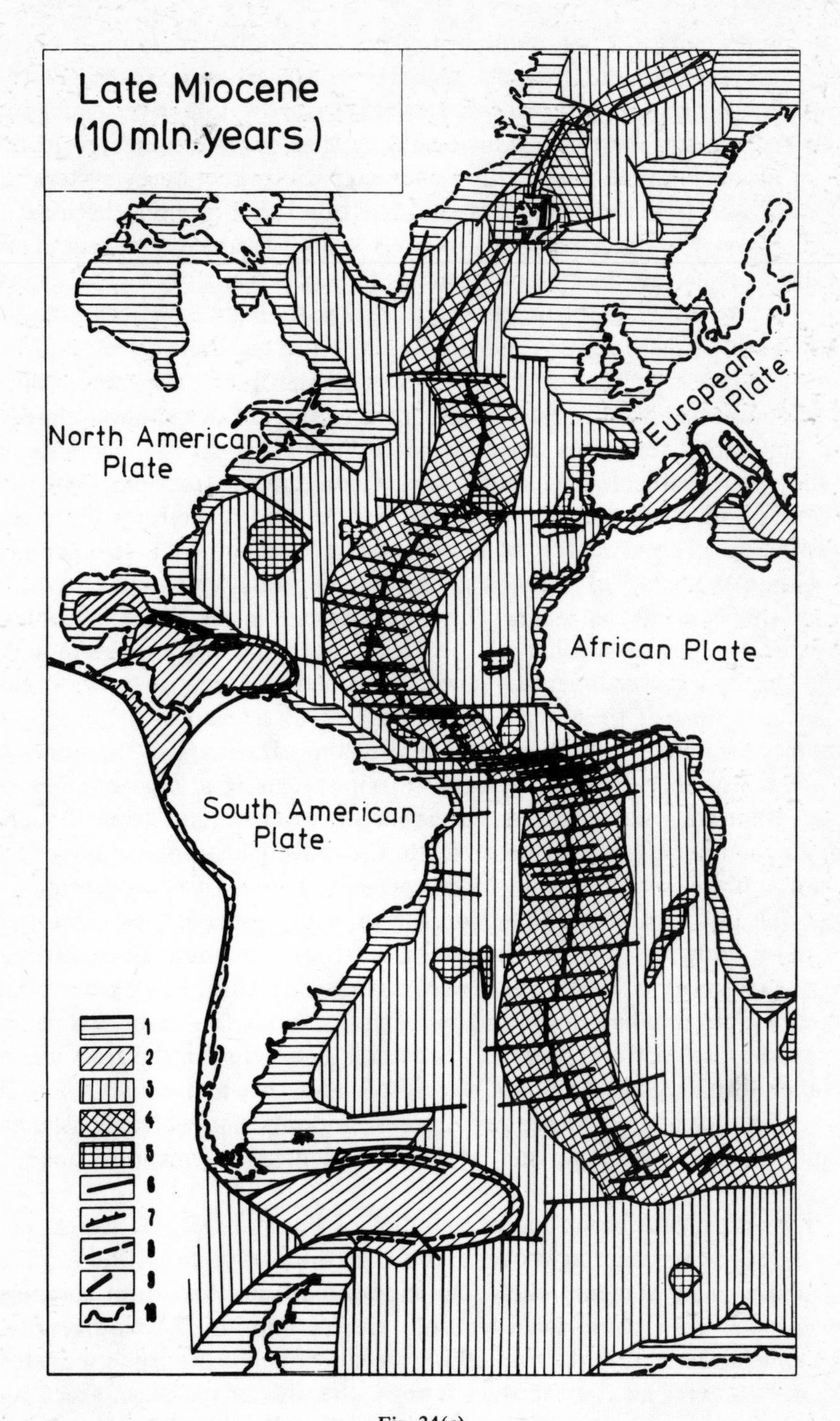
Late Miocene
(10 mln. years)
North American Plate
European Plate
African Plate
South American Plate
1
2
3
4
5
6
7
8
9
10

Fig. 34(c)

vertical movements of continental margins undoubtedly brought about the tectonic dissection of the shelf and the contental slope, subsequently modelled by exogenous processes or smoothed out by intensive sedimentation. In the Pleistocene the regions of glacial shelves were subjected to the action of glaciers which created their characteristic dissection by systems of longitudinal and transverse trenches, and the other shelf regions, because of the ocean level lowering, passed through a stage of subaerial development and valley dissection by rivers running through them. The post-glacial rise of the ocean level brought about the partial abrasive-accumulative smoothing-out of these regions and the formation of modern shelves.

Dome-and-block uplifts, ridges and swells on the floor of oceanic basins, as noted above, had been formed at the previous stage of ocean development, during the Late Cretaceous and the Early Paleogene. At the neotectonic stage they were subjected to hardly any tectonic reconstructions, with the exception of local shoves of a faulting nature, and, therefore, preserved their structure. Their surface continued being overlain by a sedimentary cover mainly made up of carbonate biogenic deposits which became quite thick by the end of the stage. This has resulted in an almost complete levelling of the primary relief of the basement. Only volcanic seamounts piercing through the sedimentary cover remained towering over the ocean floor surface. Most of these seamounts appeared as a result of the processes of oceanic basement formation in the rift zone. Having left the zone of active volcanism because of the ocean floor spreading a large number of volcanic mountains ceased their development. However, in some regions magmatic chambers were still preserved in the moving lithospheric plates for quite a long time, sometimes up to the present. This led to the appearance of large volcanic mountains found on the floor of basins and concentrated either along fault zones or in volcanic massifs. Most of these large volcanic mountains (seamounts at any rate) were extinguished in the Neogene or the Quaternary Period when they, moving together with the plate, happened to be far away from the tectonically active rift zone. But, in the areas where long-standing magmatic chambers were especially vast and active, volcanic massifs were formed, crowned by large seamounts and volcanic islands. On some of the islands, e.g., the Cape Verde Islands, the volcanoes are still active.

The formation of a block-and-range relief of the Mid-Oceanic Ridge, also traced under the sedimentary cover and on the floor of basins in the form of a rugged oceanic basement relief, has undoubtedly resulted from common causes associated with vertical tectonic shoves and the development of fissures and faults in the rift zone. With the spreading of lithospheric plates, as they were moving away from the rift zone, the process of volcano-tectonic relief formation gradually died down, although individual shoves with the formation of fault scarps, horst blocks and trench-grabens could still continue for quite a long time. Near the outer boundaries of the Mid-Oceanic Ridge the tectonic activity practically ceased, and fissures and faults closed. The only exception were the sections where volcanic activity continued; their role has been pointed out above.

In the Caribbean Transition Zone where the modern morphostructural plan started developing in the Eocene, i.e., at the neotectonic stage of development, vertical movements became manifestly predominant, although horizontal displacements on a relatively small scale are still continuing. In the Oligocene the island arc underwent an intensive uplifting, replaced in the Miocene by a period of relative quiescence, denudational levelling and deposition of limestones. But at the end of the Miocene a new phase of uplifts started, accompanied by intensive volcanism. In the south-west of the Caribbean Plate the Isthmus of Panama was formed, and it separated the Caribbean Sea from the Pacific Ocean. The Caribbean Sea basins continued to subside, which also resulted in ridges and rises. Simultaneous rise of the coastal inland regions of Central and South America led to intensive removal of sedimentary material and accumulation of turbidites in the zones of foredeeps. In the Pliocene the filling of the southern extension of the Puerto Rico Trench by sediments, which had probably started as early as the end of the Cretaceous, was completed. Because of tangential compression an inversion of the relief occurred here, and the Barbados Ridge was formed. At present the sublatitudinal branches of the Antillean Ridge are at the late orogenic stage of development, the volcanic arc of the Lesser Antilles is undergoing a transition period from the geosynclinal to the orogenic stage, and the Barbados and Curacao Ridges are only entering the late geosynclinal stage of development [99].

Exactly in the same way the history of neotectonic development can be visualized for the South Antillean Transition Zone, except that no isthmus was formed here in the rear of the Scotia Sea plate moving eastwards, and the margin of Pacific Ocean bed penetrated into the area. Sublatitudinal branches of the South Antillean Ridge underwent a multistage rise and are presently at the late orogenic stage of development, the volcanic arc of the South Sandwich Islands is entering the orogenic stage, and the floor of the Scotia Sea basin together with its uplifts and swells has been subsiding throughout the whole period.

As a result of neotectonic movements and the influence of various exogenous factors, the most important of which is sedimentation, all the diversity of large and medium-size forms of the Atlantic Ocean floor relief had been formed by the end of the Pliocene. In the Pleistocene practically no restructuring of the morphostructural plan took place, with the exception of the continuing generation of a new oceanic crust in the rift zone. More or less substantial changes in the relief are observed in the regions of active volcanism on island arcs and in volcanic massifs. On continental shelves and the summits of the largest seamounts a pronounced effect on the formation of the relief was exerted by the processes of abrasive-accumulative levelling associated with the ocean level fluctuation in the Pleistocene. Removal of terrigenous material intensified during the Pleistocene glaciations, and the correspondingly increased activity of turbidity flows led to the formation of new submarine canyons on the continental slope and the clearing of already existing ones. At the feet of continental slopes, especially in the Gulf of Mexico and on some alluvial fans in front of river mouths, thick series of Pleistocene deposits accumulated. On the ocean bed the sedimentary material is redistributed, being transported by near-bottom currents, which also penetrate into the largest and deepest trenches of the Mid-Atlantic Ridge transverse fractures, such as the Vema and Romanche Trenches. In the rift zone and on island arcs, tectonic reconstruction and active volcanism are still in progress, and this is manifested by numerous earthquake epicentres and active or comparatively recently extinguished volcanoes.

References*

1. Avdeyev, A. I., 'New map of the Caribbean Sea floor', in *Marine Hydrophysical Investigations*, No. 2, Sevastopol, 1970, No. 1, pp. 187–195.
2. Avdeyev, A. I., 'New data on the geomorphology of the Caribbean Sea Floor', in *Marine Hydrophysical Investigations*, No. 1, Sevastopol, 1973, pp. 213–225.
3. Avilov, I. K. and Gershanovich, D. Ye., 'Geomorphological investigations in the Southern Atlantic', *Izv. AN SSSR. Ser. geogr.*, 1967, No. 4, pp. 21–31.
4. Artyushkov, Ye. V., 'The origin of large stresses in the Earth's crust', *Izv. AN SSSR. Ser. Fizika Zemli*, 1972, No. 8, pp. 3–25.
5. *The Atlantic Ocean. Map. Scale 1:10000000*, Moscow: GUGK SSSR, 1971.
6. Belousov, V. V., *The Earth's Crust and the Upper Mantle of Oceans*, Moscow: Nauka, 1968, 256 pp.
7. Belousov, V. V., *Fundamentals of Geotectonics*, Moscow: Nedra, 1975, 257 pp.
8. Belousov, V. V. and Milanovsky, Ye. Ye., 'On the tectonics and tectonic position of Iceland', *Bull. MOIP, Otd. geol.*, 1975, iss. 3, pp. 81–88.
9. Beresnev, A. F., Lunarsky, G. N., Morozov, Yu. I., and Sheina L. P., 'The structure of sedimentary cover of the Mid-Atlantic Ridge from the data of continuous seismic profiling', in *Investigations on the Problem of the World Ocean Rift Zones*, v. 3, Moscow: Nauka, 1974, pp. 45–58.
10. Budanova, L. Ya., 'Relief of the Scotia Sea floor', *Trudy In-ta okeanologii AN SSSR*, 1975, **103**, 39–47.
11. Valyashko G. M., Yel'tsina G. N., Litvin V. M., Rudenko M. V., Ryabukhin A. G., Savostin L. A., and Khain V. Ye., 'Geological-geophysical characteristics of the main structural elements of the Mexican-Caribbean region', *Trudy In-ta okeanologii AN SSSR*, 1975, **100**, 54–68.
12. Gainanov, A. G., 'Isostasy and plutonic structure of the Northern Atlantic floor', *Vestnik MGU. Ser. geol.*, 1976, No. 2, pp. 39–47.
13. Gainanov, A. G. and Koryakin, Ye. D., *Geophysical Studies of the Earth's Crust Structure in the Atlantic Ocean*, Moscow: Nedra, 1967, 171 pp.
14. Gainanov, A. G. and Koryakin, Ye. D., 'Regional anomalies of gravity force in the Atlantic Ocean', in *Marine Gravimetric Investigations*, iss. 8, Moscow: MGU Publ., 1975, pp. 169–179.
15. Gerasimov, I. P., *New Approaches in Geomorphology and Paleogeography*, Moscow: Nauka, 1976, 400 pp.
16. Gerasimov, I. P., Zhivago, A. V., and Korzhuyev, S. S., 'Geomorphological and palegeographic aspects of the new theory of global plate tectonics', *Izv. AN SSSR. Ser. geogr.*, 1974, No. 5, pp. 5–22.
17. Gershanovich, D. Ye. and Dmitriyenko, A. I., 'New data on the Scotia Sea geomorphology', *Geomorfologiya*, 1972, No. 3, pp. 57–64.
18. Gorodnitsky, A. M., 'On the structure of anomalous geophysical fields over seamounts', *Okeanologiya*, 1975, **15**, 276–281.
19. Greku, R. Kh. and Avdeyev, A. I., 'Relief of the Romanche trench', in *Hydrological and Hydrochemical Investigations in the Tropical Zone of the Atlantic Ocean.*, Kiev: Naukova Dumka, 1965, pp. 136–141.

* References Nos. 1–100 are in Russian.

20. Grigoryev, S. S. and Gorodnitsky, A. M., 'Tectonics of the Azore-Gibraltar zone and its reflection in the structure of magnetic field', *Okeanologiya*, 1975, **15**, 102–107.
21. Demenetskaya, R. M., *The Crust and the Mantle of the Earth*, Moscow: Nedra, 1967, 280 pp.
22. Dmitriyev, L. V., Sharas'kin, A. Ya., Kharin, G. S., and Kurentsova, N. A., 'Petrographic characteristics of the Mid-Atlantic Ridge rift zone', in *Investigations on the Problem of the World Ocean Rift Zones*, v. 3, Moscow: Nauka, 1974, pp. 85–110.
23. Yel'nikov, I. N. and Lunarsky, G. N., 'Seismic investigations in the region of Romanche Deep and the Gulf of Guinea in the first voyage of R/V '*Akademik Kurchatov*', *Okeanologiya*, 1970, **10**, 828–836.
24. Yemelyanov, Ye. M., Lisitsin, A. P., and Ilyin, A. V. 'Types of Bottom Sediments of the Atlantic Ocean', Kaliningrad, 1975, 579 pp.
25. Zhivago, A. V., 'Geomorphology and tectonics of the South Ocean floor', in *Oceanological Studies*, No. 13, Moscow: Nauka, 1965, pp. 148–156.
26. Zhivago, A. V., Vinogradov, O. N., and Timofeyeva, N. A., 'Morphostructure of the South Ocean floor and its reflection on the new bathymetric map of Antarctica', *Izv. AN SSSR. Ser. geogr.*, 1975, No. 6, pp. 24–35.
27. Zverev, S. M., Kosminskaya, I. P., Krasil'shchikova, G. A., and Mikhota, G. G., 'Plutonic structure of Iceland and the Island-Faeroe-Shetland region from the results of seismic studies', *Bul. MOIP. Otd. geol.*, 1975, iss. 3, pp. 99–155.
28. Zonenshain, L. P. and Gorodnitsky, A. M., 'Paleooceans and the movement of continents', *Priroda*, 1976, No. 11, pp. 74–83.
29. Ivanov, M. M., '*Magnetic Survey of Oceans*', Moscow: Nauka, 1966, 183 pp.
30. Ilyin, A. V., Geomorphological investigations in the Northern Atlantic on board the R/V '*Mikhail Lomonosov*', *Trudy MGI AN SSSR*, 1960, **19**, 115–135.
31. Ilyin, A. V., 'On the question of the processes of deep ocean floor planation', *Dokl. AN SSSR*, 1963, **152**, 179–182.
32. Ilyin, A. V., 'On the morphological differences within the Mid-Atlantic Ridge', *Dokl. AN SSSR*, 1967, **172**, 913–916.
33. Ilyin, A. V., 'On the vertical movements of the Atlantic Ocean shelf in postglacial time', *Dokl. AN SSSR*, 1968, **182**, 422–425.
34. Ilyin, A. V., *Geomorphology of the Atlantic Ocean Floor*, Moscow: Nauka, 1976, 232 pp.
35. Ilyin, A. V. and Lisitsin, A. P., 'The origin of submarine canyons in connection with the peculiarities of their distribution in the Atlantic Ocean', *Dokl. AN ASSR*, 1968, **183**, 693–696.
36. Karasik, A. M., 'Magnetic anomalies of the ocean and the ocean floor spreading hypothesis', *Geotectonika*, 1971, No. 2, pp. 3–18.
37. Karasik, A. M., 'Eurasian Basin of the Arctic Ocean from the plate tectonics viewpoint', in *Problems of Geology of Polar Areas of the Earth*, Leningrad, 1974, pp. 23–31.
38. Klyonova, M. V. and Lavrov, V. M., *Geology of the Atlantic Ocean*, Moscow: Nauka, 1975, 458 pp.
39. Kogan, L. I., Korsakov, O. D., Mamayeva, N. R., Syrsky, V. N., and Greku R. Kh, 'Results of deep seismic profiling by reflection method (DSP – RWM) in the 7th voyage of R/V '*Akademik Vernadsky*', in *Integrated Geophysical Studies of the Mid-Atlantic Ridge*, Sevastopol: MGI AN USSR, 1975, pp. 45–58.
40. Kotenev, B. N. and Matishov, G. G., 'Regularities in the disjunction of the continental slope and the foot of the Northern fishery basin (the Labrador Sea, the Norwegian–Greenland Basin)', *Trudy PINRO*, 1972, iss. 28, pp. 13–22.
41. Kotenev, B. N., Nazimov, V. V., and Rvachev, V. D., 'Geomorphology of the Reykjanes Ridge', *Trudy VNIRO*, 1974, 98, 98–109.
42. Lavrov, V. M., 'Transocean fractures and their interrelation with mid-oceanic ridges and the rift system of Eastern Africa', in *Questions of Automating the Investigations of Floor Relief and the New Data on the Topography of Equatorial Atlantic*, Sevastopol: MGI AN USSR, 1969, pp. 186–196.
43. Lavrov, V. M., 'The present-day submarine volcanism of the Mid-Atlantic Ridge', *Izv. AN SSSR. Ser. geol.*, 1973, No. 2, pp. 15–24.

44. Lavrov, V. M. and Barash, M. S., 'Tectonic phases in the development of the Mid-Atlantic Ridge', *Izv. AN SSSR. Ser. geol.*, 1976, No. 3, pp. 5–12.
45. Levin, L. E., 'On the folded basement and the structure of depressions in the North and the Baltic Seas', *Izv. AN SSSR. Ser. geol.*, 1970, No. 3, pp. 70–81.
46. Leontyev, O. K., *Ocean Floor*, Moscow: Mysl', 1968. 320 pp.
47. Leontyev, O. K., 'On the changes in the world ocean level in the Meso-Cenozoic', *Okeanologiya*, 1970, **10**, 276–285.
48. Leontyev, O. K., 'On gigantic accumulative forms of the abyssal regions of the world ocean', *Okeanologiya*, 1975, **15**, 1079–1086.
49. Leontyev, O. K., 'Geodynamics of planetary morphostructures of the floor of oceans at the latest stage of the Earth's crust development', in *Fluctuations of the World Ocean Level and the Questions of Marine Geomorphology*, Moscow: Nauka, 1975, pp. 72–77.
50. Lisitsin, A. P., *Sedimentation in Oceans*, Moscow: Nauka, 1974, 438 pp.
51. Litvin, V. M., 'Relief of the Norwegian sea floor', *Trudy PINRO*, 1964, iss. 16, pp. 89–109.
52. Litvin, V. M., 'Geomorphology of the Mid-Oceanic Ridge in the Norwegian and Greenland Seas', *Okeanologiya*, 1968, **8**, 86–93.
53. Litvin, V. M., 'Relief and the bottom sediments on the shelf and continental slope at the south-eastern coast of Greenland', *Trudy PINRO*, 1970, iss. **27**, pp. 247–259.
54. Litvin, V. M., 'Geomorphology of the floor of the Norwegian and Greenland Seas', *Problemy Arktiki i Antarktiki*, 1973, iss. 42, pp. 12–16.
55. Litvin, V. M., 'Principal features of the morphostructure and tectonics of the Atlantic Ocean floor', *Izv. VGO*, 1975, **107**, 201–206.
56. Litvin, V. M., 'On the fault tectonics of the Atlantic Ocean floor', *Geotektonika*, 1975, No. 6, pp. 122–127.
57. Litvin, V. M., 'Relief and the geological structure of the Atlantic Ocean floor: Abstracts of Papers at XXIII Intern. Geogr. Cong. Section "Geography of the Ocean" ', Moscow, 1976, pp. 80–84.
58. Litvin, V. M., 'On the vertical tectonic movements of the Atlantic Ocean floor in the Meso-Cenozoic', *Okeanologiya*, 1977, **17**, 479–483.
59. Litvin, V. M. and Yemelyanova, L. P., 'The areas of the Atlantic Ocean and its parts', *Okeanologiya*, 1970, **10**, 662–669.
60. Litvin, V. M. and Rudenko, M. V., 'New data on the geomorphology of the floor in the south-eastern part of the Atlantic Ocean', *Okeanologiya*, 1971, **11**, 231–238.
61. Litvin, V. M., Marova, N. A., Rudenko, M. V., and Udintsev, G. B., 'Morphostructure of the Atlantic Ocean rift zone in the areas of the "Kurchatov" and the "Atlantis" fractures', *Okeanologiya*, 1972, **12**, 631–639.
62. Litvin, V. M. and Rudenko, M. V., 'Distribution of seamounts in the Atlantic Ocean', *Dokl. AN SSSR*, 1973, **213**, 944–947.
63. Litvin, V. M. and Sviridov, N. I., 'Seismotectonic scheme of the Atlantic Ocean floor', *Okeanologiya*, 1973, **13**, 445–450.
64. Litvin, V. M., Marova, N. A., Mirlin, Ye. G., and Udintsev, G. B., 'On the heterogeneity of the Atlantic Ocean rift zone', *Okeanologiya*, 1975, **15**, 82–88.
65. Litvin, V. M. and Yel'tsina, G. N., 'New data on the relief and the bottom sediments of the Puerto Rico and Cayman deep-sea trenches', *Okeanologiya*, 1975, **15**, 479–483.
66. Litvin, V. M., Rudenko, M. V., and Kharin, G. S., 'The role of volcanism in the formation of the Atlantic Ocean floor relief', *Geomorfologiya*, 1976, No. 4, pp. 92–98.
67. Matishov, G. G., 'Geomorphological structure of the continental slope in the Northern Atlantic', *Trudy PINRO*, 1975, iss. 35, pp. 6–15.
69. Matishov, G. G., 'The structure and origin of the marginal (longitudinal) trenches of glacial shelves', *Okeanologiya*, 1976, **16**, 259–265.
69. Mirlin, Ye. G., Nazarova, Ye. A., and Pechersky, D. M., 'Relationship between the characteristics of magnetic anomalies and the magnetic properties of Northern Atlantic basalts', *Izv. AN SSSR. Fizika Zemli*, 1975, No. 9, pp. 40–47.
70. Misharina, L. A., *Stresses in the Earth's Crust in Rift Zones*, Moscow: Nauka, 1976, 136 pp.

71. Orlyonok, V. V., 'The structure and thickness of the Atlantic Ocean sediments from seismic data', in *Oceanological Studies*, Moscow: Nauka, 1971, No. 21, pp. 271–296.
72. Orlyonok, V. V., 'Neomobilism in the light of the data on the structure of sediments on the Atlantic Ocean floor', *Geotektonika*, 1975, No. 6, pp. 111–121.
73. Orlyonok, V. V. and Gainanov, A. G., 'Geophysical investigations of the Earth's crust structure in the Labrador Sea', *Vestnik MGU. Ser. Geologiya*, 1967, No. 5, pp. 146–157.
74. Orlyonok, V. V., Litvin, V. M., Sviridov, N. I., *et al.*, 'Results of measuring the sound velocities in the bottom sediments of the Atlantic Ocean, the Baltic and the Black Seas', *Okeanologiya*, 1973, 13, 615–623.
75. Peive, A. V., 'The oceanic crust of geological past', *Geotektonika*, 1969, No. 4, pp. 5–23.
76. Peive, A. V., 'Tectonics of the Mid-Atlantic Ridge', *Geotektonika*, 1975, No. 5, pp. 3–17.
77. Pechersky, D. M. and Mirlin, Ye. G., 'A study of the nature of magnetic anomalies in the Mid-Atlantic Ridge rift zone', in *Investigations on the Problem of the World Ocean Rift Zones*, v. 3. Moscow: Nauka, 1974, pp. 129–140.
78. Popova, A. K., Suvilov, E. A., and Khobart, M., 'Geothermal studies in the Mid-Atlantic Ridge rift zone', in *Investigations on the Problem of the World Ocean Rift Zones*, v. 3. Moscow: Nauka, 1974, pp. 199–202.
79. Rvachev, V. D., 'Geomorphology of the shelf of the North-Western Atlantic', *Trudy PINRO*, 1972, iss. 28, pp. 23–47.
80. Rozhdestvensky, S. S. and Karasik, A. M., 'Linearity and symmetry in the anomalous magnetic field of the North-Atlantic Ridge crest', in *Geology of the Sea*, iss. 3, Leningrad: NIIGA, 1974, pp. 78–83.
81. Senin, Yu. M., 'Geomorphology of the western shelf of North Africa', *Trudy AtlantNIRO*, 1967, iss. 18, pp. 19–28.
82. Smirnov, Ya. B. and Popova, A. K., 'The heat flow, the age of the ocean floor, and some data for studying the driving mechanism of the tectonosphere development', *Dokl. AN SSSR*, 1975, **223**, 182–184.
83. Sorokhtin, O. G., 'Dependence of the topography of mid-oceanic ridges on the rate of the spreading of lithospheric plates', *Dokl. AN SSSR*, 1973, **208**, 1338–1341.
84. Sorokhtin, O. G., *Global Evolution of the Earth*, Moscow: Nauka, 1974, 182 pp.
85. Sorokhtin, O. G., 'Tectonics of lithospheric plates and the nature of layers of the Earth's oceanic crust', *Izv. AN SSSR. Ser. Fizika Zemli*, 1975, No. 2, pp. 50–59.
86. Syrsky, V. N., Kolezhuk, I. V., and Greku, R. Kh., 'The main features of the structure of the Vernadsky Fracture Zone at the latitude 7–8° North in the Atlantic Ocean', in *Integrated Geophysical Investigations of the Mid-Atlantic Ridge*, Sevastopol: MGI AN USSR, 1975, pp. 28–38.
87. Trifonov, V. G., 'The problems of the spreading of Iceland (the extension mechanism)', *Geotektonika*, 1976, No. 2, pp. 73–86.
88. Udintsev, G. B., 'Investigations of submarine structures in the region of Iceland' *Vestnik AN SSSR*, 1972, No. 6, pp. 82–88.
89. Udintsev, G. B., Beresnev, A. F., Verzhbitsky Ye. V., *et al.*, 'Geological-geophysical investigations in the 6th voyage of the research vessel *Akademik Kurchatov* in the Northern Atlantic', in *Structure of the Earth's Crust and Upper Mantle of Seas and Oceans*, Moscow: Nauka, 1973, pp. 3–27.
90. Udintsev, G. B., Litvin, V. M. and Kharin, G. S., 'A fire-spitting island beyond the polar circle', *Priroda*, 1974, No. 3, pp. 84–90.
91. Udintsev, G. B., Litvin, V. M., Marova, N. A., Budanova, L. Ya., and Rudenko, M. V., 'Morphostructure of the south-western part of the Walvis Ridge', *Okeanologiya*, 1976, **16**, 266–272.
92. Udintsev, G. B., Yel'nikov, I. N., Lunarsky, G. N., and Krasil'shchikova, G. A., 'Seismic observations by the refraction method on the Walvis Ridge', *Okeanologiya*, 1976, **16**, 468–472.
93. Udintsev, G. B., Litvin, V. M., and Sharas'kin, A. Ya., 'Volcanic islands in the Northern Atlantic', *Priroda*, 1976, No. 11, pp. 122–127.
94. Ushakov, S. A. and Fadeyev, V. Ye., 'On the variation of the viscous properties of the crust

and upper mantle with depth', in *Problems of the Earth's Crust and Upper Mantle Structure*, Moscow: Nauka, 1970, No. 7, pp. 32–43.

95. Fedorov, V. N., Yesyunin, R. Ye., and Grigoryeva, S. A., 'Floor relief in the tropical zone of the Atlantic Ocean', in *Questions of Automating the Investigations of Floor Relief and the New Data on the Topography of Equatorial Atlantic*, Sevastopol: MGI AN USSR, 1969, pp. 101–116.
96. Frolova, T. I., Rudnik, G. B., and Orlyonok, V. V., 'The main features of the geological development and evolution of Southern Antilles and the Scotia Sea', *Geotektonika*, 1974, No. 3, pp. 99–113.
97. Khain, V. Ye., *Regional Geotectonics*, Moscow: Nedra, 1971, 548 pp.
98. Khain, V. Ye., *General Geotectonics*. Moscow: Nedra, 1973, 510 pp.
99. Khain, V. Ye., 'Principal features of the geological development of the Mexicano-Caribbean region', *Trudy In-ta okeanologii AN SSSR*, 1975, **100**, 25–46.
100. Kholopov, B. V., Neprochnova, A. F., Lunarsky, G. N., *et al.*, 'Structure of the Earth's crust and upper mantle in the Mid-Atlantic Ridge rift zone from the data of deep-sea seismic sounding', in *Investigations on the Problem of the World Ocean Rift Zones*, v. 3, Moscow: Nauka, 1974, pp. 59–84.
101. Asmus, H. E. and Ponte, F. C., 'The Brazilian marginal basins', in *The Ocean Basins and Margins*', New York, London, 1973, Vol. 1, 'The South Atlantic'.
102. Aumento, F., Loncarevic, B. D. and Ross, D. J., 'Hudson geotraverse: geology of the Mid-Atlantic Ridge at 45° N', *Phil. Trans. Roy. Soc. London*, 1971, A **268**.
103. Avery, O. E., Burton, G. D., and Heirtzler, J. D., 'An aeromagnetic survey of the Norwegian Sea', *J. Geophys. Res.*, 1968, **73**.
104. Baker, P. E., 'Recent volcanism and magmatic variation in the Scotia Arc', in *Antarctic Geology and Geophysics*, Oslo, 1972.
105. Baker, P. E., 'Islands of the South Atlantic', in *The Ocean Basins and Margins*, New York, London, 1973, Vol. 1, 'The South Atlantic'.
106. Barazangi, M. and Dorman, J., 'World seismicity maps compiled from ESSA, Coast and Geodetic Survey, epicentre data for 1961–1967', *Bull. Seismol. Soc. Amer.*, 1969, **59**, No. 1.
107. Barker, P. F., 'Plate tectonics of the Scotia Sea region', *Nature*, 1970, **228**, No. 5278.
108. Birch, F. S., 'The Barracuda fault zone in the western North Atlantic', *Deep-Sea Res.*, 1970, **17**, No. 5.
109. Black, M., Hill, M. N., Laughton, A. S., and Matthews, D. H., 'Three non-magnetic seamounts off the Iberian coast', *Quart. J. Geol. Soc. London*, 1964, **120**.
110. Bonatti, E., 'Ancient continental mantle beneath oceanic ridges', *J. Geophys. Res.*, 1971, **76**, No. 17.
111. Bosshard, E. and McFarlane, D. J., 'Crustal structure of the Western Canary Islands from seismic refraction and gravity data', *J. Geophys. Res.*, 1970, **75**, No. 26.
112. Bott, M. H. P. and Watts, A. B., 'Deep structure of the continental margin adjacent to the British Isles', in *The Geology of East Atlantic Continental Margin. L.*, 1971, Vol. 2.
113. Bott, M. H. P., Sunderland, J., Smith P. J., *et al.*, 'Evidence for continental crust beneath the Faeroe Islands', *Nature*, 1974, **248**, No. 5445.
114. Bowin, C. O., 'Puerto Rico Trench negative gravity anomaly belt', *Geol. Soc. Amer., Mem.*, 1972, **132**.
115. Brenot, R. and Berthois, L., 'Bathymétrie du secteur Atlantique du banc Porcupine (ouest de l'Irlande) au cap Finisterre (Espagne)', *Rev. trav.*, 1962, **26**, f. 2.
116. Briseid, E. and Mascle, J., 'Structure de la marge continentale Norvégienne au debouche de la mer de Barentz', *Mar. Geophys. Res.*, 1975, **2**, 231–241.
117. Bullard, E., Everett, J. E., and Smith, A. G., 'The fit of the continents around the Atlantic', *Phil. Trans. Roy. Soc. London*, 1965, A **258**, No. 1088.
118. Bunce, E. T., 'The Puerto Rico Trench', *Geol. Surv. Canada*, 1966, Paper 66–15.
119. Bunce, E. T., Phillips, J. D., Chase, R. L., and Bowin, C. O., 'The Lesser Antilles arc and the eastern margin of the Caribbean Sea', in *The Sea*, Pergamon Press, 1970, vol. 4, pt. 2.
120. Chase, R. L. and Bunce, E. T., 'Underthrusting of the eastern margin of the Antilles by the floor of the western North Atlantic Ocean, and origin of the Barbados Ridge', *J. Geophys. Res.*, 1969, **74**, No. 6.

121. Collette, B. J., Ewing, J. I., Lagaay, R. A., and Truchan, M., 'Sediment distribution in the oceans: the Atlantic between 10° and 19°N', *Mar. Geol.*, 1969, **7**, No. 4.
122. Collette, B. J., Schouten, H., Rutten, K., and Slootweg, A. P., 'Structure of the Mid-Atlantic Ridge province between 12° and 18°N', *Mar. Geophys. Res.*, 1974, **2**, 143.
123. Cox, A., Doell, R. R., and Dalrymple, G. B., 'Reversals of the Earth's magnetic field', *Science*, 1964, **144**, 1537–1543.
124. Dickson, G. O., Pitman, W. C., and Heirtzler, J. R., 'Magnetic anomalies in the South Atlantic and sea-floor spreading', *J. Geophys. Res.*, 1968, **73**, No. 6.
125. Drake, C. L., Campbell, N. J., Sander, G., and Nafe, J. E., 'A Mid-Labrador Sea Ridge', *Nature*, 1963, **200**, No. 4911.
126. Eldholm, O. and Windisch, C. C., 'Sediment distribution in the Norwegian – Greenland Sea', *Bull. Geol. Soc. Amer.*, 1974, **85**, No. 11.
127. Emery, K. O., Uchupi, E., Phillips, J. D. *et al.*, 'Continental rise off Eastern North America', *Bull. Amer. Assoc. Petrol. Geol.*, 1970, **54**, No. 1.
128. Emery, K. O., Uchupi, E., Bowin, C. O., *et al.*, 'Continental margin off Western Africa: Cape St Francis (South Africa) to Walvis Ridge (South-West Africa)', *Bull. Amer. Assoc. Petrol. Geol.*, 1975, **59**, No. 1.
129. Emery, K. O., Uchupi, E., Phillips, J., *et al.*, 'Continental margin off Western Africa: Angola to Sierra Leone', *Bull. Amer. Assoc. Petrol. Geol.*, 1975, **59**, No. 12.
130. Engel, A. E. J. and Engel, C. G., 'Mafic and ultramafic rocks', in *The Sea*, Pergamon Press, 1970, vol. 4, pt. 1.
131. Ewing, J., 'Seismic model of the Atlantic Ocean', in *Earth's Crust and Upper Mantle*, Washington, 1969.
132. Ewing, J. and Ewing, M., 'Seismic-refraction measurements in the Atlantic Ocean basins, in the Mediterranean Sea, on the Mid-Atlantic Ridge, and in the Norwegian Sea', *Bull. Geol. Soc. Amer.*, 1959, **70**, No. 3.
133. Ewing, J. I., Edgar, N. T., and Antoine, J. W., 'Structure of the Gulf of Mexico and Caribbean Sea', in *The Sea*, Pergamon Press, 1970, vol. 4, pt. 2.
134. Ewing, J. I., Ludwig, W. J., Ewing, M., and Eittreim, S. L., 'Structure of Scotia Sea and Falkland Plateau', *J. Geophys. Res.*, 1971, **76**, No. 29.
135. Ewing, M. and Heezen, B. C., 'Puerto Rico Trench topographic and geophysical data', *Geol. Soc. Amer., Spec. Pap.*, 1955, **62**.
136. Ewing, M., Le Pichon, X., and Ewing, J., 'Crustal structure of the mid-ocean ridges. 4: Sediment distribution in South Atlantic Ocean and the Cenozoic history of the Mid-Atlantic Ridge', *J. Geophys. Res.*, 1966, **71**, No. 6.
137. Ewing, M., Carpenter, G., Windisch, C., and Ewing, J., 'Sediment distribution in the oceans: the Atlantic', *Bull. Geol. Soc. Amer.*, 1973, **84**, No. 1.
138. Fainstein, R., Milliman, J. D., and Jost, H., 'Magnetic character of the Brazilian continental shelf and upper slope', *Rev. Brasil. Geosci.*, 1975, **5**, No. 3.
139. Fox, P. J., Lowrie, A. J., and Heezen, B. C., 'Oceanographer Fracture Zone', *Deep-Sea Res.*, 1969, **16**, No. 1.
140. Fox, P. J., Ruddiman, W. F., Ryan, W. B. F., and Heezen, B. C., 'The geology of the Caribbean crust: Beata Ridge', *Tectonophysics*, 1970, **10**, 495–513.
141. Fox, P. J., Schreiber, E., and Heezen, B. C., 'The geology of the Caribbean crust: Tertiary sediments, granitic and basic rocks from the Aves Ridge', *Tectonophysics*, 1971, **12**, No. 2.
142. Francheteau, J. and Le Pichon, X., 'Marginal fracture zones as structural framework of continental margins in South Atlantic Ocean', *Bull. Amer. Assoc. Petrol. Geol.*, 1972, **56**, No. 6.
143. Frank, S. and Nairn, A. E. M., 'The equatorial marginal basins of West Africa', in *The Ocean Basins and Margins*, New York, London, 1973. Vol. 1, 'The South Atlantic'.
144. Fransis, T. J. G., 'Upper mantle structure along the axis of the Mid-Atlantic Ridge near Iceland', *Geophys. J. Roy. Astron. Soc.*, 1969, **17**, No. 5.
145. Freeland, G. L. and Dietz, R. S., 'Plate tectonics evolution of Caribbean-Gulf of Mexico region', *Nature*, 1971, **232**, No. 5305.
146. Goslin, J. and Sibuet, J. C., 'Geophysical study of the easternmost Walvis Ridge, South Atlantic: deep structure', *Bull. Geol. Soc. Amer.*, 1975, **86**, No. 12.

147. Green, D. H. and Ringwood, A. E., 'The genesis of basaltic magma', *Contrib. Mineral. and Petrol.*, 1967, **15**, 103–190.
148. Hammond, A. L., 'Project FAMOUS: exploring the Mid-Atlantic Ridge', *Science*, 1975, 187, No. 4179.
149. Harbison, R. N., Lattimore, R. K., and Rona, P. A., 'Structural lineations in the Carnary Basin, eastern central North Atlantic', *Mar. Geol.*, 1973, **14**, No. 4.
150. Hayes, D. W. and Ewing, M., 'North Brazilian Ridge and adjacent continental margin', *Bull. Amer. Assoc. Petrol. Geol.*, 1970, **54**, 2120–2150.
151. Heezen, B. C., Tharp, M., and Ewing, M., 'The floor of the oceans. 1: The North Atlantic', *Geol. Soc. Amer. Spec. Pap.*, 1959, **65**.
152. Heezen, B. C. and Tharp, M., *Physiographic diagram of the South Atlantic Ocean, the Caribbean Sea, the Scotia Sea, and the eastern margin of the South Pacific Ocean*, N.Y.: Geol. Soc. Amer., 1961.
153. Heezen, B. C., Bunce, E. T., Hersey, J. B., and Tharp, M., 'Chain and Romanche Fracture Zones', *Deep-Sea Res.*, 1964, **11**, No. 1.
154. Heezen, B. C. and Johnson, G. L., 'The South Sandwich trench', *Deep-Sea Res.*, 1965, **12**, No. 2.
155. Heezen, B. C. and Tharp, M., *Physiographic diagram of the North Atlantic Ocean*. N.Y.: Geol. Soc. Amer., 1968.
156. Heezen, B. C., Johnson, G. L., and Heirtzler, C. D., 'The northwest Atlantic mid-ocean canyon', *Canad. J. Earth Sci.*, 1969, **6**.
157. Heirtzler, J. R. and Hayes, D. E., 'Magnetic boundaries in the North Atlantic Ocean', *Science*, 1967, **157**, No. 3785.
158. Heirtzler, J. R., Dickson, G. O., Herron, E. M., *et al.*, 'Marine magnetic anomalies, geomagnetic field reversals and motions of the ocean floor', *J. Geophys. Res.*, 1968, 73, No. 6.
159. Heirtzler, J. R. and Bryan, W. B., 'The floor of the Mid-Atlantic rift', *Sci. Amer.*, 1975, **233**, No. 2.
160. Hekinian, R., 'Volcanics from the Walvis Ridge', *Nature*, 1972, **239**, 91–93.
161. Hekinian, R. and Aumento, F., 'Rocks from the Gibbs Fracture Zone and the Minia Seamount near 53° N in the Atlantic Ocean', *Mar. Geol.*, 1973, **14**, No. 1.
162. Hinz K. and Moe, A., 'Crustal structure in the Norwegian Sea', *Nature*, 1971, **232**, No. 35.
163. Holtedahl, H., 'Some remarks on geomorphology of continental shelves off Norway, Labrador and southeast Alaska', *J. Geol.*, 1958, **66**, No. 4.
164. Hood, P., 'Magnetic surveys of the continental shelves of Eastern Canada', *Geol. Surv. Canada, Pap.*, 1966, **66–15**.
165. *Initial Reports of the Deep-Sea Drilling Project*, Wash.: Nat. Sci. Found., 1969–1976. Vols. 1–4, 10–15, 36–39.
166. Isacks, B., Oliver, J., and Sykes, L. R., 'Seismology and new global tectonics', *J. Geophys. Res.*, 1968, **73**, No. 18.
167. Johnson, G. L. and Heezen, B. C., 'Morphology and evolution of the Norwegian-Greenland Sea', *Deep-Sea Res.*, 1967, **14**, No. 6.
168. Johnson, G. L. and Vogt, P. R., 'Morphology of Bermuda Rise', *Deep-Sea Res.*, 1971, **18**, No. 6.
169. Johnson, G. L., Vogt, P. R., and Schneider, E. D., 'Morphology of the Northeastern Atlantic and Labrador Sea', *Dtsch. Hydrogr., Ztschr.*, 1971, **24**, No. 2.
170. Johnson, G. L., Southall, J. R., Young, P. W., and Vogt, P. R., 'Origin and structure of the Iceland Plateau and Kolbeinsey Ridge', *J. Geophys. Res.*, 1972, 77, No. 29.
171. Johnson, G. L., Lowrie, A., and Hey, R. N., 'Marine geology in the environs of Bouvet Island and the South Atlantic triple junction', *Mar. Geophys. Res.*, 1973, **2**, 23.
172. Kearey, P., Peter, G., and Westbrook, G. K., 'Geophysical maps of the Eastern Caribbean', *J. Geol. Soc.*, 1975, **131**, No. 3.
173. Keen C. and Tramontini, C., 'A seismic refraction survey on the Mid-Atlantic Ridge', *Geophys. J. Roy. Astron. Soc.*, 1970, **20**.
174. Keen, M. J., 'Possible edge effect to explain magnetic anomalies off the eastern seabord of the U. S.', *Nature*, 1969, **222**, No. 5188.

175. Kharin, G. S., Litvin, V. M., and Rudenko, M. V., 'Volcanic rocks and their role in the bottom structure of the Atlantic Ocean', in *Volcanoes and tectonosphere*, Tokai Univ. Press, 1976.
176. Krause, D. C., 'Guinea Fracture Zone in the Equatorial Atlantic', *Science*, 1964, **146**, No. 3640.
177. Larson, E. E. and La Fountain, L., 'Timing of the breakup of the continents around the Atlantic as determined by paleomagnetism', *Earth Planet. Sci. Lett*., 1970, **8**, No. 5.
178. Laughton, A. S., 'South Labrador Sea and the evolution of the North Atlantic', *Nature*, 1971, **232**, No. 5313.
179. Laughton, A. S., Whitmarsh, R. B., Rusby, J. S. M., *et al*., 'A continuous east-west fault on the Azores – Gibraltar Ridge', *Nature*, 1972, **237**, No. 5352.
180. Laughton, A. S., Roberts, D. G., and Graves, R., 'Bathymetry of the Northeast Atlantic: Mid-Atlantic Ridge to Southwest Europe', *Deep-Sea Res*., 1975, **22**, No. 12.
181. Le Pichon, X., 'Sea-floor spreading and continental drift', *J. Geophys. Res*., 1968, **73**, No. 12.
182. Le Pichon, X. and Langseth, M. G., 'Heat flow from the mid-ocean ridges and sea-floor spreading', *Tectonophysics*, 1969, **8**, 319–344.
183. Le Pichon, X. and Hayes, D. E., 'Marginal offsets, fracture zones and the early opening of the South Atlantic', *J. Geophys. Res*., 1971, **76**, No. 26.
184. Le Pichon, X., Francheteau, J., and Bonnin, J., *Plate Tectonics*. Amsterdam, New York: Elsevier, 1973.
185. Leyden, R., Ludwig, W., and Ewing, M., 'Structure of the continental margin off Punta del Este, Uruguay, and Rio de Janeiro, Brazil', *Bull. Amer. Assoc. Petrol. Geol*., 1971, **55**, No. 12.
186. Lonardi, A. G. and Ewing, M., 'Sediment transport and distribution in the Argentine Basin. 4: Bathymetry of the continental margin, Argentine Basin and other related provinces', in *Physics and Chemistry of the Earth*, Pergamon Press, 1971, vol. 8.
187. Loncarevic, B. D., Mason, C. S., and Matthews, D. H., 'Mid-Atlantic Ridge near 45° North. 1: The Median Valley', *Canad. J. Earth Sci*., 1966, **3**.
188. Ludwig, W. J., Ewing, J. I., and Ewing, M., 'Structure of Argentine continental margin', *Bull. Amer. Assoc. Petrol. Geol*., 1968, **52**, No. 12.
189. Machens, E., 'The geology history of the marginal basins along the north shore of the Gulf of Guinea', in *The Ocean Basins and Margins*, New York, London, 1973, Vol. 1, 'The South Atlantic'.
190. Markl, R. G., Bryan, G. M., and Ewing, J. I., 'Structure of the Blake-Bahama outer ridge', *J. Geophys. Res*., 1970, **75**, No. 24.
191. Martin, H., 'The Atlantic margin of Southern Africa between latitude 17° South and the Cape of Good Hope', in *The Ocean Basins and Margins*, New York, London, 1973, Vol. 1, 'The South Atlantic'.
192. Mayer, P., 'Voläufige Ergebnisse magnetisch-bathymetrischen Messungen im Nordatlantik zwischen 41°N und 44°N', *Ztschr. geol. Wiss*., 1976, **4**, No. 4.
193. McGregor, B. A. and Rona, P. A., 'Crest of the Mid-Atlantic Ridge at 26°N', *J. Geophys. Res*., 1975, **80**, No. 23.
194. McMaster, R. L., Lachance, T. P., Asharf, A., and Boer, J. de., 'Geomorphology, structure and sediments of the continental shelf and upper slope off Portuguese Guinea, Guinea, and Sierra Leone', in *The Geology of East Atlantic Continental Margin*, London, 1971, Vol. 4.
195. Melson, W. G., Hart, S. R., and Thompson, G., 'St. Paul's rocks, Equatorial Atlantic: petrogenesis, radiometric ages, and implications on sea-floor spreading', *Geol. Soc. Amer., Mem*., 1982, **132**.
196. Meyer, O., Voppel, D., Fleischer, U., *et al*., 'Results of bathymetric, magnetic and gravimetric measurements between Iceland and 70°N', *Dtsch. Hydrogr. Ztschr*., 1972, **25**, No. 5.
197. Meyerhoff, A. A. and Hatten, C. W., 'Bahama Salient of North America', in *The Geology of Continental Margins*, New York, Berlin, 1974.
198. Miyashiro, A., 'Metamorphism and related magmatism in plate tectonics', *Amer. J. Sci*., 1972, **272**, 629–656.
199. Miyashiro, A., Shido, F., and Ewing, M., 'Petrologic models for the Mid-Atlantic Ridge', *Deep-Sea Res*., 1970, **17**, No. 2.
200. Molnar, P. and Oliver, J., 'Lateral variations of attenuation in the upper mantle and discontinuities in the lithosphere', *J. Geophys. Res*., 1969, **74**, No. 10.

201. Montadert, L., Winnock, E., Delteil, J. -R., and Grau, G., 'Continental margins of Galicia-Portugal and Bay of Biscay', in *The Geology of Continental Margins*, New York, Berlin, 1974.
202. Nicholls, G. D., Nalwalk, A. J., and Hays, E. E., 'The nature and composition of rock samples dredged from the Mid-Atlantic Ridge between 22° N and 52° N', *Mar. Geol.*, 1964, **1**, No. 4.
203. Olivet, J. –L., Le Pichon, X., Monti, S., and Sichler, B., 'Charlie-Gibbs fracture zone', *J. Geophys. Res.*, 1974, **79**, No. 14.
204. Pitman, W. C. and Talwani, M., 'Sea-floor spreading in the North Atlantic', *Bull. Geol. Sci. Amer.*, 1972, **83**, No. 3.
205. Pitman, W. C., Larson, R. L. and Herron, E. M., *Magnetic Lineations of the Oceans* (chart), N.Y.: Lamont-Doherty Geol. Obser., 1974.
206. Pratt, R. M., 'Great Meteor Seamount', *Deep-Sea Res.*, 1963, **10**, No. 1/2.
207. Roberts, D. G., 'New geophysical evidence on the origin of the Rockall plateau and trough', *Deep-Sea Res.*, 1971, **18**, No. 3.
208. Rona, P. A., 'Relation between rates of sediment accumulation on continental shelves, sea-floor spreading, and eustasy inferred from the Central North Atlantic', *Bull. Geol. Soc. Amer.*, 1973, **84**, No. 9.
209. Rona, P. A., Harbison, R. N., and Bush, S. A., Abyssal hills of the Eastern Central North Atlantic', *Mar. Geol.*, 1974, **16**, No. 5.
210. Saito, T., and Funnell, B. M., 'Pre-quaternary sediments and microfossils in the oceans', in *The Sea*, Pergamon Press, 1970, Vol. 4, pt. 1.
211. Sclater, J. G. and Detrick, R., 'Elevation of mid-ocean ridges and the basement age of Joides Deep-Sea Drilling sites', *Bull. Geol. Soc. Amet.*, 1973, **84**, No. 5.
212. Sclater, J. G. and McKenzie, D. P. 'Paleobathymetry of the South Atlantic', *Bull. Geol. Soc. Amer.*, 1973, **84**, No. 10.
213. Searle, R. C., 'Lithospheric structure of the Azores Plateau from Rayleigh-wave dispersion', *Geophys. J. Astron. Roy. Soc.*, 1976, **44**, No. 3.
214. Shepard, F. and Emery, K. O., 'Congo submarine canyon and fan valley', *Bull. Amer. Assoc. Petrol. Geol.*, 1973, **57**, No. 9.
215. Sheridan, R. E., 'Atlantic continental margin of North America', in *The Geology of Continental Margins*, N.Y., Berlin, 1974.
216. Sheridan, R. E., Houtz, R. E., Drake, C. L., and Ewing, M., 'Structure of continental margin off Sierra Leone, West Africa', *J. Geophys. Res.*, 1969, **74**, No. 10.
217. Sykes, L. R., 'Mechanism of earthquakes and nature of faulting on the Mid-Oceanic Ridge', *J. Geophys. Res.*, 1967, **72**, No. 8.
218. Sykes, L. and Ewing, M., 'The seismicity of the Caribbean region', *J. Geophys. Res.*, 1965, **70**, 5065–5074.
219. Talwani, M. and Le Pichon, X., 'Gravity field over the Atlantic Ocean', *Geophys. Monogr.* 13, Amer. Geophys. Union, 1969.
220. Talwani, M., Le Pichon, X., and Ewing, M., 'Crustal structure of the mid-ocean ridges. 2: Computed model from gravity and seismic refraction data', *J. Geophys. Res.*, 1965, **70**, No. 2.
221. Talwani, M., Windisch, C. C., and Langseth, M. G. 'Reykjanes Ridge Crest: A Detailed Geophysical Study', *J. Geophys. Res.*, 1971, **76**, No. 2.
222. Thorarinsson, S., 'The median zone of Iceland', in *The World Rift System*, Ottawa, 1966.
223. Uchupi, E., 'Atlantic continental shelf and slope of the United States. Physiography', U. S. Geol. Surv. Prof. Pap., 1968, 529–C.
224. Uchupi, E., Phillips, J. D., and Prada, K. E., 'Origin and structure of the New England Seamount chain', *Deep-Sea Res.*, 1970, **17**, No. 3.
225. Uchupi, E., Emery, K. O., Bowin, C. O., and Phillips, J. D., 'Continental margin off Western Africa: Senegal to Portugal', *Bull. Amer. Assoc. Petrol. Geol.*, 1976, **60**, No. 5.
226. Ulrich, J., 'Zur Topographie des Reykjanes-Rückens', *Kiel. Meeresforsch.*, 1960, **16**, No. 2.
227. Ulrich, J., 'Geomorphologische Untersuchungen an Tiefsee-Kuppen im nordatlantischen Ozean', *Verh. Dtsch. Geographentag*, 1970, **37**.
228. Van Andel, T. H. and Bowin, C. O., 'The Mid-Atlantic Ridge between 22° N and 23° N latitude and tectonics of mid-ocean rises', *J. Geophys. Res.*, 1968, **73**, No. 4.

229. Van Andel, T. H. and Heath, G. R., 'Tectonics of the Mid-Atlantic Ridge, 6–8° South latitude', *Mar. Geophys. Res.*, 1970, **1**, 5.
230. Van Andel, T. H., von Herzen, R. P., and Phillips, J. D., 'The Vema Fracture Zone and the tectonics of transverse shear zones in oceanic crustal plates', *Mar. Geophys. Res.*, 1971, **1**, 261.
231. Van Andel, T. H., Rea, D. K., Von Herzen, R. P., and Hoskins, H., 'Ascension Fracture Zone, Ascension Island, and the Mid-Atlantic Ridge', *Bull. Geol. Soc. Amer.*, 1973, **84**, No. 5.
232. Vine, F. J. and Matthews, D. H., 'Magnetic anomalies over oceanic ridges', *Nature*, 1963, **199**, No. 4897.
233. Vogt, P. R., 'Volcano height and plate thickness', *Earth Planet. Sci. Lett.*, 1974, **23**, 337–348.
234. Vogt, P. R., Schneider, E. D., and Johnson, G. L., 'The crust and upper mantle beneath the sea', *Geophys. Monogr. 13*. Amer. Geophys. Union, 1969.
235. Vogt, P. R., Anderson, C. N., and Bracey, D. R., 'Mesozoic magnetic anomalies, sea-floor spreading, and geomagnetic reversals in the southwestern North Atlantic', *J. Geophys. Res.*, 1971, **76**, No. 20.
236. Whitmarsh, R. B., 'Axial intrusion zone beneath the median valley of the Mid-Atlantic Ridge at 37° N detected by explosion seismology', *Geophys. J. Roy. Astron. Soc.*, 1975, **42**, No. 1.
237. Wilson, J. T., 'Evidence from islands on the spreading of ocean floor', *Nature*, 1963, **197**, No. 4867.
238. Wilson, J. T., 'A new class of faults and their bearing on continental drift', *Nature*, 1965, **207**, No. 4995.

Subject Index

Geographical Index